VEGANS SAVE THE WORLD

Plant-based Recipes and Inspired Ideas for Every Week of the Year

ALICIA ALVREZ

Published by Mango Publishing Group, a division of Mango Media Inc.

Cover Image : VICUSCHKA/shutterstock.com
Cover + Layout Design : Elina Diaz

For permission requests, please contact the publisher at:

Mango Publishing Group
2850 Douglas Road, 3rd Floor
Coral Gables, FL 33134 USA
info@mango.bz

For special orders, quantity sales, course adoptions and corporate sales, please email the publisher at sales@mango.bz. For trade and wholesale sales, please contact Ingram Publisher Services at customer.service@ingramcontent.com or +1.800.509.4887.

Vegans Save the World: Plant-based Recipes and Inspired Ideas for Every Week of the Year

Library of Congress Cataloging
ISBN: (paperback) 978-1-63353-657-9, (ebook) 978-1-63353-658-6
Library of Congress Control Number: 2017911897
BISAC category code: CKB125000 COOKING/Vegan

Printed in the United States of America

"Veganism is not a 'sacrifice.' It is a joy."

– Gary L. Francione

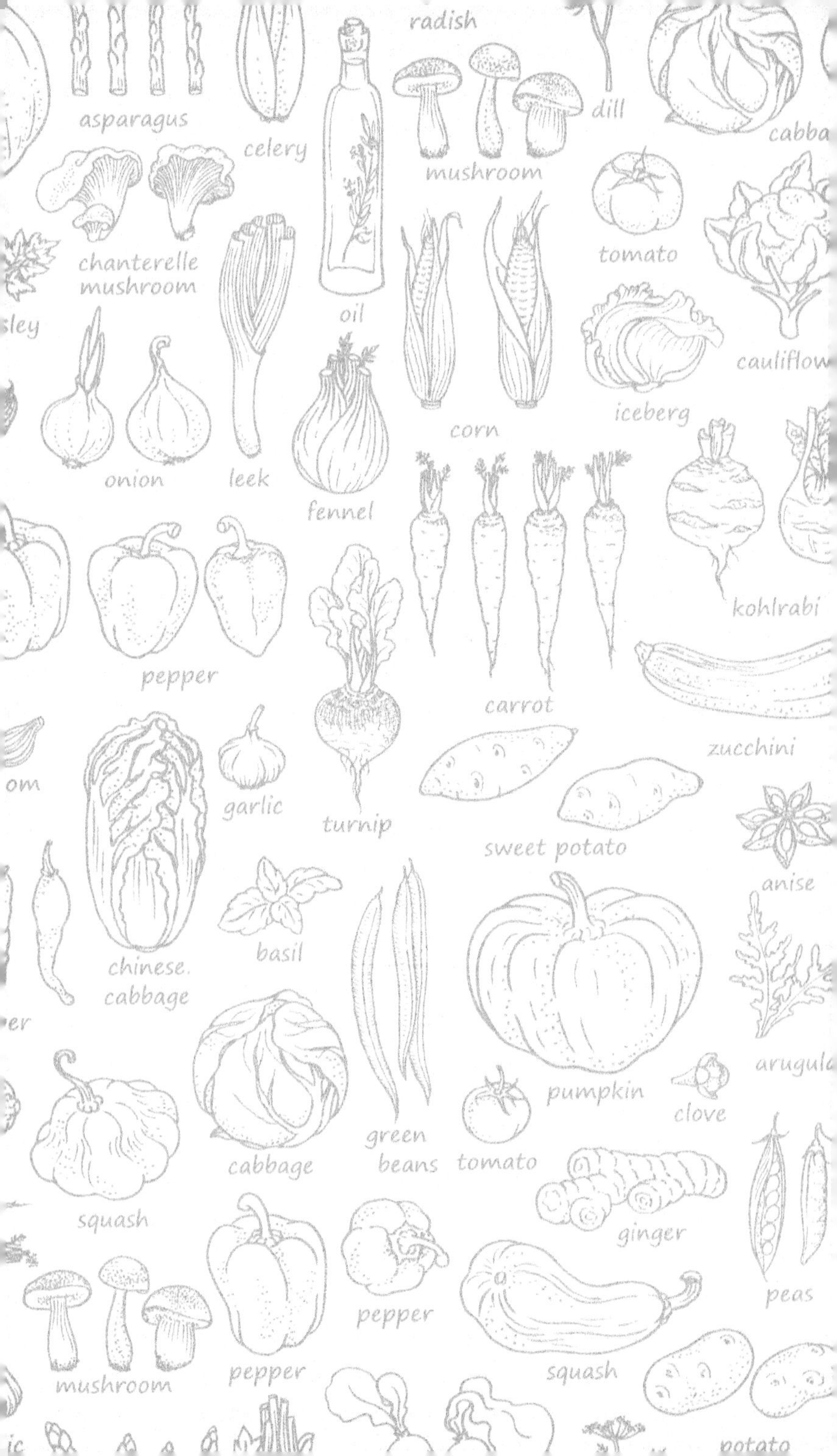
radish
asparagus
celery
mushroom
dill
chanterelle mushroom
tomato
oil
iceberg
corn
onion
leek
fennel
kohlrabi
pepper
carrot
zucchini
garlic
turnip
sweet potato
anise
chinese cabbage
basil
pumpkin
clove
green beans
tomato
cabbage
ginger
squash
peas
pepper
pepper
squash
mushroom
potato

Contents

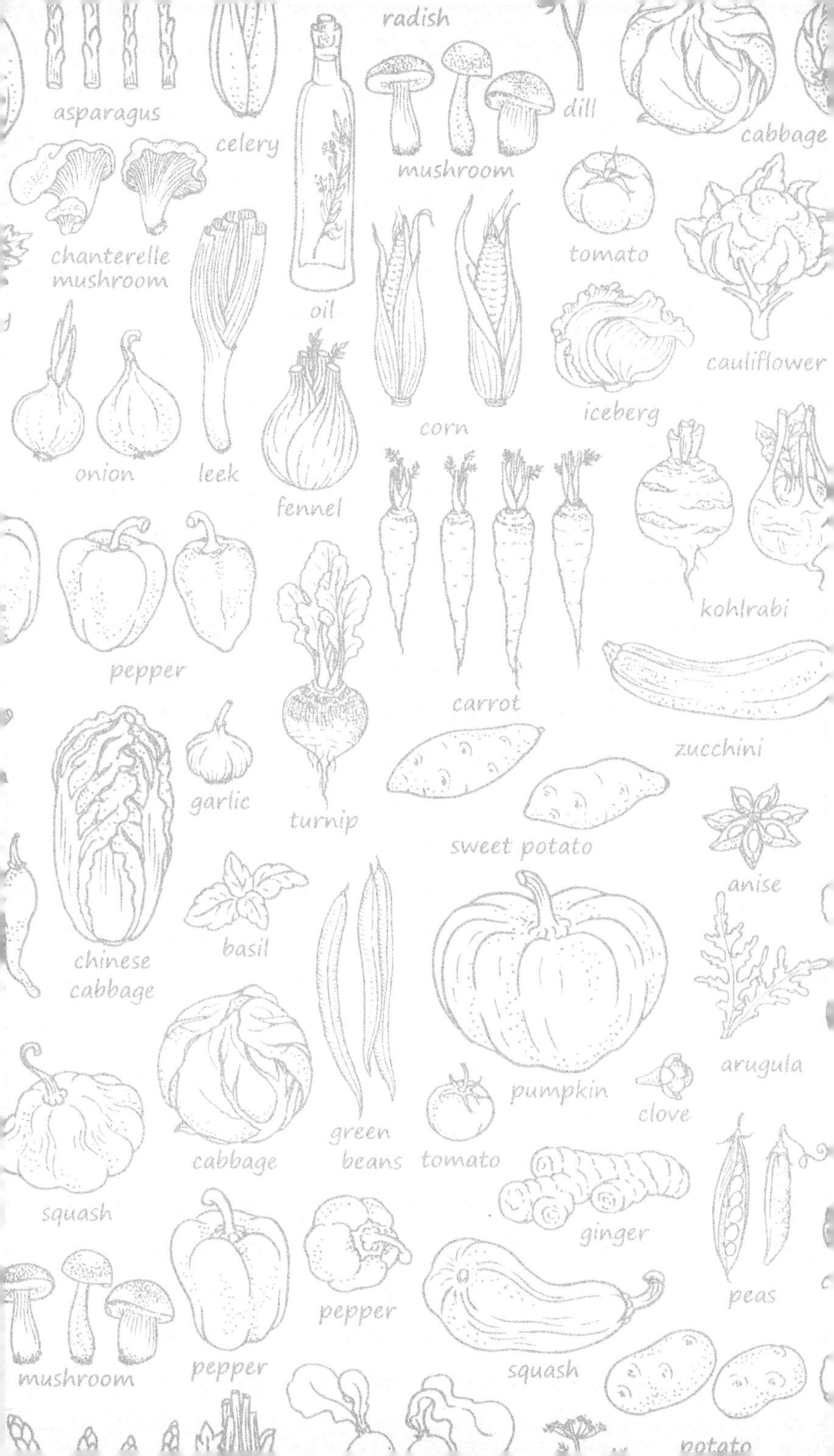

radish
asparagus
celery
mushroom
dill
cabbage
chanterelle mushroom
tomato
oil
cauliflower
iceberg
corn
onion
leek
fennel
kohlrabi
pepper
carrot
zucchini
garlic
turnip
sweet potato
anise
chinese cabbage
basil
arugula
pumpkin
clove
green beans
tomato
cabbage
squash
ginger
peas
pepper
mushroom
pepper
squash

radish
asparagus
celery
mushroom
dill
chanterelle mushroom
tomato
oil
corn
iceberg
onion
leek
fennel
kohlrabi
pepper
carrot
zucchini
garlic
turnip
sweet potato
anise
chinese cabbage
basil
arugula
pumpkin
clove
green beans
tomato
cabbage
squash
ginger
peas
pepper
mushroom
pepper
squash
potato

Foreword

I love this plant-based lifestyle cookbook in part because I can truly relate to this approach to food and the importance of feeding your family in a healthful and sustainable way. I also consider myself to be vegan, living the good life as a result. I was raised in a "meat and potatoes" household, but I have turned to a more vegan diet partly because of food sensitivities such as dairy. Lucky for me, others in my household are lifelong vegetarians and have fantastic recipes. They are over-the-moon for the yummy menus and inspired ideas in *Vegans Save the World.*

Any cook can use this book, even if not totally vegan or vegetarian. If you want to focus on creating a healthy diet for your loved ones that is predominantly whole foods and plant-based, this is the perfect companion for that deliciously healthy journey!

The recipes are fairly easy to prepare and are delicious. Best of all, they got the all-important approval from my grade-school kids. The recipes and variations are written in a logical progression of steps, which is great and might be extra helpful for someone just learning to cook with lots of veggies and grains. There are many thoughtful suggestions that definitely enliven the vegan or vegetarian diet to expand your repertoire of plant-based meals, including ideas for special occasions. My favorite aspect of this guide to getting healthy is all the options and encouragement to follow your own desires as to the specific ingredients or to use what you might have on hand or what is in season. This is a cookbook I love and highly recommend to others. I keep it in my kitchen to refer to often, continuing to turn to it over and over again.

Join us for the adventure of plant-based living. Eating vegan is good for you, good for your family and contributes to a sustainable, planet-positive lifestyle.

How vegans can save the world:

Eating for the Environment

Cutting back on meat consumption is good for the environment, your health and your wallet. Designate one day a week "meat free" and know that you are helping reduce greenhouse gas emissions. Producing one pound of beef puts as much carbon dioxide into the environment as driving a typical car 70 miles! Read "Livestock's Long Shadow," the 2006 UN paper on the effects of the meat industry on the environment and human populations, to learn more.

Just Say No to GMO

Buy organic heritage seedlings whenever you can. Most fruits and vegetables have an incredibly diverse range of varieties, but we typically only see one or two different kinds in the grocery store. By choosing to grow heritage plants we can preserve that diversity and give a big, green thumbs-down to monoculture.

More Beef = Fewer Trees

The next time you consider grabbing a burger at a fast food restaurant, remember this: Over the past few decades, the rainforests have been disappearing to satisfy our "hunger" for cheap beef. Rainforests are home to over a thousand indigenous tribal groups and thousands of species of birds, butterflies and exotic animals—all of which are now endangered. Rainforests also affect rainfall and wind all around the world by absorbing solar energy for the circulation of our atmosphere. The trees provide buffers against wind damage and soil erosion, which then helps prevent flooding along our coastlines. They are a precious part of our ecosystem. Let's all do something to protect them.

Over five million acres of South and Central American rainforests are cleared each year for cattle to graze on. The local people don't eat this much meat—it is exported to make the $1 hamburger and a cheap barbeque meal in America.

Save Energy, Clean Air

Not only do livestock farms produce carbon dioxide and other harmful gases, but producing animal-based protein also requires more fossil-fuel energy than plant-based. This means that the meat industry not only pollutes the air, but uses an incredible amount of energy too. By not eating meat, or even eating less meat, we could contribute to a cleaner atmosphere and a healthier environment by not adding to fossil fuel use or carbon dioxide creation.

Help Feed More People

Right now, most grain that is grown in the United States goes to feed livestock. Cornell did a study and found that with the amount of grain that goes to feed livestock, 800 million people could be fed. Additionally, land that could be used for farming is being destroyed by livestock farms that don't look after the soil. Being conscious of not eating meat could change this system to allow more people to be fed.

It's More Ethical

Most of us are no strangers to the conditions lots of livestock animals live in before heading to the slaughterhouse. Packing animals into tight spaces, not allowing them to see sunlight or grass, from birth until death is a cruel way of existence. By buying cheap, factory-produced meat, or going to fast food restaurants that use factory-produced meat, you contribute to this cycle of horrible conditions.

Sir Paul McCartney and many other celebrities support Meat Free Mondays. Check it out at www.meatfreemondays.co.uk. The Belgian city of Ghent has instituted a Meat Free Thursday. Get inspired and start a local meat free day! Start a farm-to-school project in your school district; all the know-how is at www.farmtoschool.org. Making small steps like these go a long way in helping us keep our planet happy, and ourselves healthy.

Level 1

Vegan Basics

How to Get Started

TIP

Knowing other vegans can help with getting support about becoming vegan. Use social media, message boards or Meetup to get in contact with other people so you don't feel so alone.

How to Get Started

You've read about what being a vegan means and you're ready to take the plunge. It can be a daunting process so here are the first few steps to help you get going.

Do Your Research: You've already done some reading, but do you really know what being vegan is all about? Obviously, this book is a fantastic step to introduce you to the ins and outs of veganism. You probably shouldn't make a huge life change after reading just one book. Read a few, along with magazine articles and online blogs.

Take it Slow: Some people say that stopping all animal products in one fell swoop is the way to go (going cold turkey, if you'll forgive the phrase). But it can be really hard when you consider the huge range of things you are suddenly going to have to cut out of your diet and your life.

A good first step is to get rid of meat. Once you've started to master cooking vegetarian meals, start to let go of eggs and dairy. After that, you can start to really examine labels and ferret out all the hidden animal ingredients in your food. Even if it takes months to get to a real vegan place, it will be worth it.

Learn to Cook: If you don't already have great cooking skills, now's the time to learn. So many conventional and processed foods are non-vegan that you'll almost certainly have to start doing a lot more home-cooking to make it work.

Try New Foods: It's not just about what you take out of your diet. Make up a batch of quinoa, do a stir fry with tempeh or grill up a black bean burger. Experimentation will open up so many new doors that you won't even realize that some things are gone. Try at least one new food a week.

You're Not Perfect: Once you start to notice all the animal products in so many things around you, it can be pretty overwhelming. Don't throw your hands up and quit because you find your hand lotion has some glycerin in it. Being as vegan as possible is better than just giving up.

The specifics of how you get started aren't that important. What matters is that you do it.

Isn't a Vegan Diet Expensive?

This isn't really an easy question to answer because it's not as cut and dry as you might think. It's a mixture of new things you will want to buy, as well as some things you won't be buying anymore. How that final equation balances out will be different for everyone.

According to NPR, the average American household spends around 21% of its food budget on meat, and then another 11% on dairy products. That means approximately one third of your grocery budget is going to animal products. Once you start a vegan diet, all that money then gets freed up to be used for other purchases. And since meat is considerably more expensive than fruits, grains, and vegetables to begin with, that means you can buy a lot more food while not changing your overall expenditures.

So buying that organic soy milk may seem like an added expense compared to the usual cow's milk, but don't forget that you won't be buying any chicken breasts, pork tenderloin or roasting beef either. You have to look at the whole picture before you can really calculate the costs.

The trick is not to fall into the replacement trap. As a new vegan, you'll quickly find that you spend most of your money on the meat, dairy and egg replacements. You have to shift your eating habits altogether to add more fruits, grains, nuts and vegetables to each meal rather than find some "fancy" alternative to the meat you are used to. That's how you keep your budget under control.

The bottom line is that many vegan products can cost more than their conventional counterparts, but when you factor in all the meat and dairy you're not eating, it isn't as bad as you might think. With a little planning and effort, you probably can eat for less as a vegan, even with that organic soy milk or non-dairy cheese.

Isn't a Vegan Diet Expensive?

TIP

According to NPR, we spend **23% of our food budgets** on processed foods and snacks. Avoid these to cut costs—many contain animal products anyway.

Look for the Logo

TIP

If your food product has this symbol, it meets the criteria listed above. Look for it at the grocery store!

Look for the Logo

One way to speed up your shopping is to keep your eyes open for the vegan symbols on your food products. It sure is simpler than trying to read all the ingredients for yourself.

Unfortunately, there isn't one single regulating body for a universal vegan logo. The most common symbol comes from the Vegan Society, which has been advocating for vegan living since 1944. In order to register for their trademark, a product has to:

- **Contain or involve no animals, animal products, by- products or derivatives**
- **Have no involvement in animal testing during development or manufacture**
- **Contain no GMOs that have animal-derived genes. Plant-derived genes are acceptable but the products must be clearly labelled as GMO**
- **Cross-contamination with non-vegan materials is kept to a minimum**

And to be clear, their definition of "animal" includes insects, invertebrates and any other beings that would be classed in the scientific Animalia kingdom (cite below).

You can find the Vegan Society logo on vegan products around the world. It's a registered symbol in the USA, Canada, Australia, and across Europe covering more than 16,000 products.

When a product doesn't have a vegan symbol, it doesn't mean the food is automatically non-vegan. It just means the manufacturer hasn't gotten themselves registered. Reading the ingredients can still be a fine way to determine what you're eating. It just takes longer.

History of Modern Veganism

Interested in this whole veganism idea and want to know how it all started? Here's a little history about the movement.

The best place to find the origins of veganism is with the early history of the Vegan Society, the first organized group of vegans. It was founded in 1944, meaning that modern vegans have been around for more than 70 years. It's not quite as recent a "fad" as some people think.

The founders include Donald Watson, Elsie Shrigley, and other vegetarian friends. They were discussing the idea of being vegetarian while also eliminating all dairy products from their diet. The term "vegan" was coined, taking the first and last few letters from vegetarian. Donald Watson felt it represented the beginning and end of conventional vegetarianism.

By 1949, this little group had expanded and they officially declared their purpose to be more than about diet, but the complete freedom of animals from being used by man. They formed as a registered charity in 1979, and have been going strong ever since.

The Current State of Veganism:

According to a survey done for the Vegetarian Resource Group (cite below), there are approximately 16 million vegetarians in the USA, and they estimate that about half of those are fully vegan. So it may seem like nobody else you know is vegan, but there are millions who have given up animal products completely.

As for the Vegan Society, their symbol is the standard trademark for vegan products around the world and their website is the main online destination for resources, information, and product details for vegans.

History of Modern Veganism

TIP

Joining the Vegan Society yourself **can help you get some connections** as well as supporting their cause.

How Veganism Helps Animal Welfare

TIP

Avoid brands of products that are tested on animals, like makeup & shampoo. You can stop contributing to

systemic animal cruelty, one purchase at a time.

How Veganism Helps Animal Welfare

This is the core issue of being vegan: the treatment and welfare of animals (all animals, including insects).

What distinguishes vegans from vegetarians is that we recognize that there is more to animal treatment than simply not killing them. People get very hung up on the idea that taking eggs or milk isn't harm, and that it's just fine. That's a very myopic view of things, and not accurate at all. In fact, animals kept for wool, milk, or eggs are probably suffering more than those kept for meat because they suffer for many more years.

Since you are already reading a book about veganism, you're probably very aware of animal welfare issues around a traditional meat-eating diet. But if you're still on-the-fence about the seriousness of this global problem, keep reading.

Not all animals undergo the same cruelties, but there are some pretty consistent issues across the board in the livestock industry. In almost any case, animals are kept in extremely small spaces living in filth. They are treated cruelly with no concern for their pain or fear. Antibiotics are used almost constantly because livestock would surely die from infection from their living conditions otherwise.

For the "lucky" animals not destined for slaughter, life is not good. Eggs come from chickens kept in tiny cramped cages for their entire lives, and dairy cows undergo repeated pregnancies and constant milking for years until they die an exhausted death. In most dairy operations, male calves aren't needed so they are sent off to be sold as veal. That means their short lives are spent locked in containers so they can't move, to keep their muscles tender.

Sheep kept for wool don't fare any better, though this tends to be a surprise to most people. Careless shearing conditions lead to cuts, scrapes and other bloody injuries that are not usually treated properly.

The situations aren't just about crowding, disease and dirty conditions either. Staff in many of these livestock facilities treat the animals horribly. There are countless videos that have been taken showing beatings and other violent abuse that goes on in factory farms.

The only way to eliminate all of this abuse is to remove the consumer demand for these products. If people stopped buying meat and other animal products, there would be no need to hold billions of livestock animals captive. That's how being a vegan helps.

How Veganism Helps the Environment

Being vegan is mainly about the ethics of using animals for our benefit, but there are other important reasons why you might want to give up eating animal products. Besides helping the animals themselves, a vegan diet can help the environment as a whole.

Raising animals for food is a huge drain on the environment, and it uses up many more resources that using the same space to produce plant-based food.

To start with, keeping animals for meat is an enormous waste of water, which is a resource we are starting to run out of in many parts of the world. It will take more than 2,000 gallons of water to eventually produce 1 pound of beef for consumption. A pound of wheat would only require 25 gallons. Add in the use of fossil fuel to maintain the vast amounts of crops for feed as well as everything else (transport, slaughter, and processing). On average, it will use up 11 times as much fuel to produce one calorie of animal protein as it would for an equal calorie of plant protein.

Now, that's just wasted resources. The bigger shame comes in land use. Not only are millions of acres used for pasture and feed lots, even more land is taken up in the growing of feed crops. Food is grown, just not for us. According to PETA, approximately one third of the entire planet's land surface is used to house livestock animals or grow their feed.

The problem isn't just that space is wasted and native ecosystems are razed, but land is made dead and barren by overgrazing and the repeated growing of the same crops. Land used as pasture gets heavily contaminated with animal manure, which can negatively impact the local water supplies too. The EPA estimates that 35,000 miles of rivers in 22 states are polluted due to animal waste runoff.

The bottom line is that the animal food industry is a horrible drain on the Earth's space and resources. We can all eat much better with less environmental damage if we ditch the animals.

TIP

A lot of major food companies that make fruits & veggies aren't environmentally friendly, either. **Buy local or grow your own** to reduce your impact on the Earth.

How Veganism Helps Your Body

TIP

Create a varied & healthy diet for yourself. Eating nothing but kale and soy burgers may be vegan but it's not going to do your body any favors.

How Veganism Helps Your Body

A vegan lifestyle isn't just extremely good for the world, it's personally good for your own body. For all the concerns about protein and other nutrients, a plant-based diet is a healthy one. You'll do yourself a lot of good by saying good bye to meat and other animal products.

Cholesterol: We all know about cholesterol, a type of fat that is only found in animal products. Too much of it can lead to hardening arteries and heart disease. Since cholesterol doesn't exist in the plant world, this is one health problem you won't have to worry about.

Other forms of fat are pretty high in meat as well, though you can get high-fat vegan foods too (avocados are one example). Overall, you'll find your fat intake drops considerably when you ditch the animal foods.

Antibiotics: These are unfortunately becoming a common "ingredient" in most meat and dairy products raised in factory conditions. Animals are given huge amounts of antibiotics to keep them from getting sick in their dirty living conditions, and those antibiotics are still present in the meat or dairy products once they hit the shelves. They are in the environment as well as your body, and it's slowly but surely helping to create antibiotic resistant "superbugs," or illnesses that cannot be treated because they have mutated to be unaffected by the drugs.

Hormones: Along with antibiotics, hormones are also being added to meat products which means you're consuming them when you eat meat (or eggs or dairy). They're used to force animals to grow larger or faster than what is natural, or to mature earlier. That makes livestock farming more profitable, even if it's very unhealthy. When you eat meat or milk products, you're consuming traces of these hormones which can lead to a disruption in your own body chemistry.

Vitamins, Fiber & Antioxidants: The health benefits of being vegan aren't just from the things you're avoiding, but also the increase in your intake of nutrients. The variety of fruits, vegetables, beans, nuts and seeds that are part of a vegan diet offer much more for your body than meat products do. By filling up your plate with healthier plant-based food, you're going to be getting more nutrition overall. If you're still worried about protein, check out the section on vegan proteins. It's not a problem.

Celebrity Vegans

By no means are we saying that you should make a major lifestyle choice because some movie star has done it. But sometimes it can be reassuring to see big names making the same choices you are, which can give you a sense of being connected to the larger vegan community.

Also, a list like this gives you the opportunity to support projects involving other vegans.

- Ariana Grande
- Carrie Underwood
- Kristen Bell
- Bryan Adams
- Tobey Maguire
- Ellen DeGeneres
- Jared Leto
- Samuel L. Jackson
- Eliza Dushku
- Ellen Page
- Jessica Chastain
- Joan Jett
- Laura Prepon
- Johnny Galecki
- Sarah Silverman
- Mayim Bialik
- Russell Brand
- Peter Dinklage
- Paul McCartney

Many celebrity vegans are quite outspoken about their lifestyle and often work towards animal welfare and environmental causes. This list is based on statements made by the celebrities, and is as accurate as possible. There is no way of knowing when or if they change their eating habits, or if they are being totally up front in the first place.

Celebrity Vegans

TIP

Follow outspoken vegan celebrities on social media to get updates on news or campaigns related to animal welfare *AND* the environment.

TIP

Vegan restaurants are increasing in number.
If you're going out with friends, suggest some vegan cuisine.
Even simply decreasing meat consumption is a big help.

Can You Make a Difference?

Once you start reading about the millions of animals being held in such poor conditions in the food industry, it can be overwhelming.

Well, one person cannot change the entire industry—that much is clear. But for every person who does, it's one step closer. You'd be surprised how a single person's lifestyle and diet can impact the lives of hundreds of animals.

To put it in perspective, here are some figures on what the "average" person consumes in a year:

- **189 pounds** of liquid milk per year (about 23 gallons)
- **31 pounds** of cheese
- **200 pounds** of pork, beef, and poultry
- **260** eggs

Granted, it's not that easy to visualize the actual animals behind these figures, but you can still get the overall idea. For the 200 lbs of meat, that roughly represents 1 entire pig. So over a decade of vegan eating, you are "saving" the lives of 10 adult pigs. And that's only regarding meat.

In just a year, that's a lot of animal-based food you're NOT eating. And don't forget, these are just the obvious products that come directly from animals. By avoiding all the hidden ingredients as well as animal-based materials like leather, you are adding to your personal impact.

And you can have more of an impact on the cruelty of meat-eating with actions beyond your personal choices. Sign petitions, write letters, and donate to causes. Because of activism, the world-renowned Ringling Brothers circus will retire their elephants, and public pressure is changing the rules about keeping chickens in battery cages.

The more people request meat-less options in restaurants, the more likely it is that restaurants adjust their menus. The more stores are asked to carry vegan products, the more options we're all going to have. And they'll start to get cheaper with increased demand. It all adds up.

Slowly but surely, people can and do make a difference.

Level 2

Foods & Products to Avoid

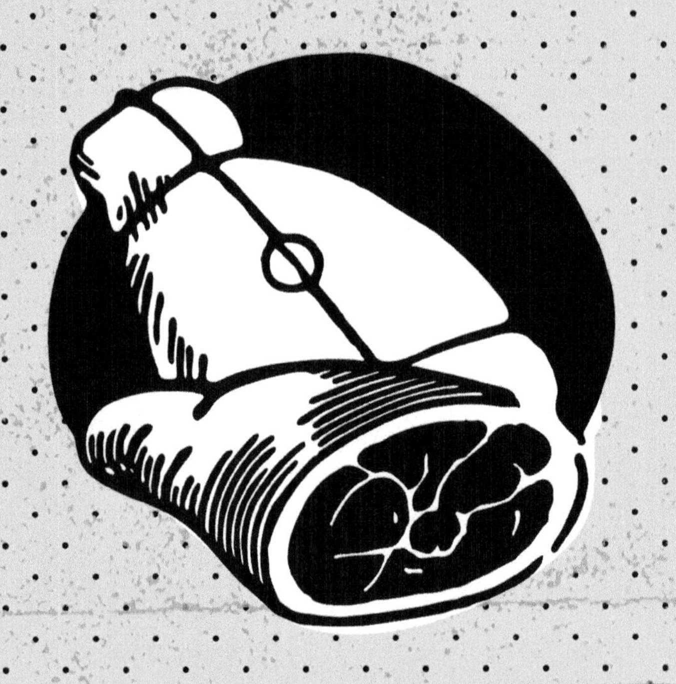

Avoiding Meat

TIP

Look up vegetable-based recipes online if you're short on ideas for a yummy dish. Don't get bored just eating steamed broccoli

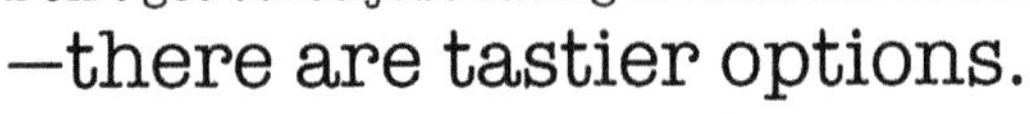
—there are tastier options.

Avoiding Meat

Here is the central portion of the vegan diet, the exclusion of meat. In many ways, it's actually the simplest to figure out. It's not the kind of product you usually find sneaking around in ingredient lists so you can take care of your meat consumption with a little less investigation. That doesn't mean that it's an easy thing to eliminate from your diet.

We're talking about any kind of animal flesh, so that means beef, pork, chicken, turkey, duck, goat, sheep, rabbit, wild game as well as fish and other seafood. Other animal products like dairy and eggs are covered in their own sections later in the book.

Commercial Substitutes: There is a big market for meat substitutes, and you can buy "meat" strips for stir fries, patties for burgers and crumbles for any dish requiring ground beef. They are mainly made from soy, and they aren't always necessarily vegan. Some may have egg or other animal-based product in them, making them fine for vegetarians but not vegans. Just read the ingredients before choosing a brand.

Vegetable-based Proteins: Again, there is another section for this because there are 4 or 5 different ways you can get your daily protein from vegetable based sources. Tofu, tempeh and seitan are just a few of them and you'll want to get to know how to cook them.

Don't Replace At All: This is the toughest approach because the standard North American diet is so used to meals with a large portion of meat. There is no reason why you need to start worrying about new kinds of vegan meat. Let it go entirely, and you'll find life simpler. Beans and nuts can bring you the protein you need, and you don't need to design your meals to revolve around a "meat" main course. Just serve pasta, rice, vegetables, or beans in any combination without the focus on meat.

Other Animal Ingredients: While actual meat itself isn't usually hidden in other products, there are a few animal-based items you have to watch out for. Gelatin is made from boiled down cartilage, bones and tendons, and lard is a common form of pig fat. Both of these are addressed in other sections as hidden ingredients in various other foods you have to watch out for.

Avoiding Milk

Unlike meat, milk is produced without any slaughter of the animals (usually cows, but sometimes goats or sheep are kept commercially for milk too). It's still an animal welfare issue though. Female cows are kept constantly pregnant so they produce milk, and the calves are taken away immediately and sold off as meat animals. Most dairy cows only survive 4 years (instead of the natural 25-year lifespan) because of the harsh conditions.

That Darned Calcium: One of the biggest myths about milk is that it's a necessary part of a healthy diet because of all the wonderful calcium it provides. The truth is that milk isn't really a great source at all. It's just conventional wisdom. If your diet includes several of these foods on a regular basis, you'll get enough calcium:

- **kale**
- **molasses**
- **figs**
- **white beans**
- **black-eyed peas**
- **spinach**
- **turnip greens**
- **seaweed**
- **sesame seeds**

You can easily eat a healthy diet without worrying about having to replace milk. Once you stop needing milk so much, it's easier to replace it when you do need to.

Milk Replacements: Even if you're getting your nutrients, having milk around is kind of important for most baking and lots of other delicious recipes where it's a required ingredient. Soy milk works well for cooking and drinking, but you can try one of the newer brands of almond, rice, coconut or even cashew milk.

Hidden Milk: Not only do you have to avoid milk as a product itself, but as an ingredient. And it's not always milk itself you need to watch for. Hidden milk by-products such as casein, lactose, and whey are all non-vegan too. Whey is particularly frustrating because it is in so many foods.

Avoiding Milk

TIP

Foods that may contain hidden milk ingredients include boxed cereals AND cereal bars, prepared breadcrumbs, granola, and protein or energy bars.

Avoiding Dairy Products

TIP

Look for vegan ice cream recipes online! You don't have to give up your favorite peanut butter chunk or triple chocolate desserts to be cruelty-free.

Avoiding Dairy Products

We already looked into removing milk itself from your diet, along with some alternatives you can use. But there are so many other foods made from milk that the discussion really doesn't stop there. What about ice cream, sour cream, butter, and all the different varieties of cheese?

Ice Cream: Try one of the many frozen desserts made with rice milk or soy milk. There are several brands out there. Check out Almond Dream or So Delicious. For an easy DIY approach, whip up a few frozen bananas in a blender with some fruit, and you'll have some quick and simple non-dairy "ice cream."

Sour Cream: Tofutti makes a pretty good non-dairy sour cream product, though it isn't always the easiest to find except in health food stores or really large supermarkets. Make an easy homemade vegan alternative with a mix of soaked cashew nuts in a blender with a little lemon juice.

Butter: Don't be fooled by the idea you can replace butter with standard supermarket margarine. Even though they are considered to be non-dairy products, they usually have lactose, casein and/or whey in them. These are all still milk by-products, making margarine non-vegan. For an actual vegan alternative, try Earth Balance brand.

Alternatively, replace small amounts of butter in baking with equal amounts applesauce or mashed avocado. For a half cup of butter or more, you'll want proper margarine though. For frying, try a little olive oil instead of butter.

Cheese: Vegan cheese has been a tough product to really develop. One of the best is Daiya, who makes cheddar and mozzarella shreds that have great taste as well as that unique cheese texture.

Another product you can use to add a cheesy flavor to your cooking is nutritional yeast. There is another section on it specifically where you can get the details. Add to sauces or sprinkle on other foods like grated parmesan. For cream cheese, you can make your own with similar whipped nuts as the sour cream, but with some additional straining to thicken up the texture.

Avoiding Fish

Many vegetarians ease away from meat by switching to fish. As a vegan, now it's time to give that up completely.

Fish are killed by dragging them out of the water to slowly suffocate, and that goes for wild-caught as well as farmed fish. Farmed fish may help keep wild populations from being completed destroyed, but the fish are kept in dirty and overcrowded conditions for their short lives.

Hidden Fish: Avoiding fish as a meat ingredient isn't that hard. There are a few instances where hidden fish by-products can sneak into your food though. Alcoholic drinks are often filtered with isinglass, made from dried fish bladders, and traditional Worcestershire sauce is flavored with anchovies.

You also have to watch for foods with added vitamin supplements. Vitamins A and D3 are often derived from fish oils, and the popular omega-3 fatty acids you see added to so many products are also often sourced from fish (but not always). Orange juice is one particular culprit. Check your ingredient lists to see which ones have fish-derived vitamins. If you are looking to get more omega-3s, stick with flax or hemp seeds. They are far better sources anyway.

Replacing Fish: Unlike most other meat products, there really isn't much of a market for "fish replacements". The brand Gardein does make a yummy vegan "fishless fillet" that tastes like battered fish (like you'd have with fish and chips). One notable exception is tuna salad. A common vegan replacement for this tasty sandwich filling can be made with cooked chickpeas, vegan mayo, chopped onion, celery and a little dill. Recipes are easy to find online. If you're just looking for those fatty acids, add more flax to your diet.

Avoiding Fish

TIP

If you want to "veganize" a seafood dish, think about texture as well as taste. **If you want a chewy, flaky texture** (like for fillets or fish sticks), **tempeh is a good choice.**

Avoiding Eggs

TIP

If your main concern is animal cruelty and you'd like to keep eggs in your diet, **consider raising your own chickens.** You'll have a new pet and a guilt-free egg supply.

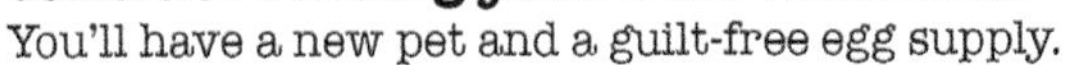

Avoiding Eggs

Though the idea of "free range" eggs is starting to be better known in the general population, the reality is that most chickens are kept in battery cages. They live their lives in a space no bigger than a sheet of computer paper. In the USA alone, there are more than 300 million chickens kept like this. So even though the collecting of eggs doesn't directly harm chickens, the industry that abuses them does.

Replacing Eggs: First, you'll want a few tips on how to replace eggs in your cooking. That is, dishes where you actually taste the eggs (e.g. not baking). The usual go-to for egg replacement is tofu. Chop up some firm tofu, and scramble it up with a few spices and veggies and you have a vegan version of scrambled eggs.

Eggs in Baking: Getting rid of eggs in your diet can be particularly frustrating when it comes to baking. The unique binding properties of eggs means you can't just leave them out. You have a few choices to add that glue to your recipes without the egg.

Flax seeds create a sticky gel when soaked in water, and it makes a fantastic replacement for eggs in baking. You'll need 1 tablespoon of ground flax and 3 tablespoons of water. Mix it up in a small dish, and wait for half an hour. You can also use 1/4 cup of whipped soft tofu, soy yogurt or a mashed banana to replace an egg in baking.

For a store-bought option, try Ener-g egg replacement. It might work better in any baking that needs to be really fluffy or light.

What About Free Range? You can't really talk about eggs without mentioning the new trend of free range. Sure, we'd like to visualize large grassy pastures of happy chickens pecking at passing insects. That's not even close to being true. Most free range chickens are kept in just as crowded conditions, but not confined to small cages. Thousands can be kept in sheds, so overpopulated you can't see the floor.

Watch Out for Hidden Eggs: Actually, eggs and related products aren't as sneaky as some others. While you will definitely find eggs as an ingredient is lots of foods (namely baked products), it's usually indicated clearly as "egg" in the ingredients. Albumin is another name for egg white that is sometimes listed. Also you might want to watch for vitamin A and lecithin. Both of these compounds can be extracted from eggs, but they may come from other sources. Check with the manufacturers to find out for sure.

The Happy Hour Vegan

This might be a bit of a shock, but many alcoholic beverages and cocktails contain animal products, or were processed with animal products at some point. So before you indulge in that glass of wine, find out what's in it.

The Actual Ingredients: Though alcohol doesn't carry ingredient labels like regular food or drinks do, you can usually find out the ingredients through the brewer or just from general information about what goes into certain drinks.

Meat isn't too likely to show up in your drink, but there are many classic beverages made with honey or dairy. Mead is completely made with honey, and there are several kinds of beer flavored with honey as well. The beers are usually easy to spot because their name reflects the honey content. Benedictine also contains honey, which is a little less obvious.

Creamy liqueurs are harder to figure out. Unlike the situation with honey, many beverages have the term "cream" in their name even though they don't actually contain any dairy. So, you'll have to do a bit more research on that. Irish Cream under various brands is a very well-known liqueur that does contain dairy, so that's one place to start.

The Filtering Process: Here is where it gets more difficult. After fermentation most beverages (particularly beer and wine) are cloudy, and not that appealing to look at. Unfortunately, many traditional filtering techniques involve a wide (and slightly strange) range of animal products. Some of them include:

- **Gelatin** - a protein derived from tendons and bones of animals
- **Isinglass** - another kind of collagen that comes from dried fish air bladders
- **Albumen** - protein that comes from egg whites
- **Bone char** - one kind of filtering charcoal is made from animal bones
- **Casein** - a protein from milk

These are definitely not vegan options. Don't despair. There are methods of filtering that are vegan-friendly, such as with clay, diatomaceaous earth and even some natural forms of moss are used. Some small breweries skip the filtering altogether since it's mainly for looks anyway.

The Happy Hour Vegan

TIP

For beer, major brands like Budweiser, Coors Molson **are typically vegan.** A large database of drinks can be searched for vegan-status at Barnivore.com.

TIP

If you're hesitant about giving up honey but want to help bees, look into new technologies like Flow Hive, which propose to make collecting honey a stress-free process for bees.

Free the Bees

Everyone thinks about the usual animal products when it comes to going vegan. You automatically think of meat, eggs and dairy. Honey often comes as a surprise when people start thinking about the sources of their food. Sure, bees are "just" insects, but it still falls under the same umbrella of animal exploitation.

The National Honey Board estimates that there are a more than 2 and a half million bee colonies in the United States alone (not including any hobbyists who have fewer than 5 hives) (cite below). With an average population of around 50,000 insects per hive, that adds up to more than 100 billion bees held in captivity for our use.

Though bees are not directly killed in order to harvest their honey, many die simply by accident while the keepers smoke them and take the hives apart for access to the honey-filled combs. Even if none are killed or harmed, vegans don't feel that it's right for humans to take the honey that the bees worked so hard to produce.

Watching Out for Honey: Avoiding actual honey as a stand-alone product is easy enough so that's not a problem. It can also be used as an ingredient, so start reading labels. Food products that are trying to get away from processed sugar have added honey as a sweetener instead, usually touting themselves as healthier. Look out for baked products, granola bars and some boxed cereals. Several brands of alcoholic beverages are also made with honey, particularly any kind of mead.

Replacing Honey: You'll have to stay away from processed sugar because it's not vegan either (we'll cover that in another chapter), though you can safely use raw or unrefined sugars as a sweetener. For the same liquid consistency, you can also try agave nectar, golden corn syrup or maple syrup. These are all a lot like honey but is plant-sourced and completely vegan. The maple syrup will give a different flavor though.

What About Pollination? Just like with other vegan concerns, there are some hidden ones here that you may not see. Some hives are kept and maintained in order to provide pollination for fruit and vegetable gardening. Most vegans are OK with this as long as the bees are basically left alone to live their lives and their honey not collected. Trying to avoid any fruits or vegetables that may have been pollinated by a "captive" bee colony is nearly impossible anyway.

Hidden Ingredients in Sweets

TIP

Research the kind of sugar you're buying. The use of pesticides, for example, is one way in which farming sugarcane is harmful to the environment.

Hidden Ingredients in Sweets

Sweet treats can easily hide many non-vegan ingredients, especially some that you wouldn't even recognize if you read them. As we look at the unexpected animal products in food, here are some of the ingredients to look out for in candies and other such goodies.

Sugar: Yes, that's right. Any sort of refined or processed sugar (both white and brown!) are considered to be non-vegan. It's not the sugar itself, but the refining process.

Bone char is a kind of charcoal that is derived from animal bones, and it's used in production to make sugar all sparkly white. To complicate things, commercial brown sugar is actually white sugar with molasses added back to it. It's still non-vegan. Sugar can be refined in a vegan manner, it's just not as common. Check your local stores for raw or organic sugar. These will be made without the bone char, making them vegan-friendly.

Confectioner's Glaze: It's also sometimes called confectionery shellac, and it comes from an insect. The lac beetle secretes this resin, and it's harvested by the ton for the food industry. On average, it takes 300,000 beetles to make 1 kg of shellac resin. The bugs themselves are not harvested, rather the resin is crusted on tree branches and harvested from there. Unfortunately, a lot of insects die in the process because it is just scraped off without clearing away the beetles first.

Some shiny coatings are made with carnauba wax, which would be fine for vegans.

Some Colorings: Most candies get their vibrant coloring from a myriad of chemicals, usually all artificial. But anything with the terms cochineal or carmine is made from crushed beetles.

Gelatin: Though the animals aren't directly killed for gelatin, it's most definitely an animal product. Marshmallows and Jello are the biggest culprits for gelatin, but it's also a common ingredient in frosting, lots of gummy candies, and frosted cereals.

Chocolate: Thankfully, pure chocolate can be vegan. It's the various milk chocolate varieties you have to watch out for. There are specifically vegan chocolate products on the market, but any dark or even semi-sweet variety can be vegan.

Hidden Ingredients in Snacks

Other snack foods also hide a mix of animal by-products that you can easily miss. It's not as common as in the sweets we just talked about, but there are a few things you need to be aware of.

Potato Chips: The flavorings used in many brands of potato chips (and other similar crunchy snacks) are often less vegan than you'd think. Pretty much anything cheese-flavored will have some combination of dairy by-products, and some even have forms of chicken fat (particularly BBQ flavors). Some deep-fried snacks like these are made with lard, a type of rendered animal fat.

Cakes Mixes: You can also find lard in many brands of cake mixes, particularly those that don't require eggs to be added. Packaged snack cakes can also have lard in them, with Hostess brand being the most notorious for this. Of course, many cake products are already non-vegan due to the milk or egg ingredients in them. Still, it's good to know what else may be lurking in there.

Snack Peanuts: You can't get much simpler than a jar of peanuts. Most people probably wouldn't even bother to read the ingredient list. If you do, you'll notice that some brands use gelatin. As we mentioned earlier, gelatin is made from boiling down bones, ligaments, hooves, and other unwanted animal body parts. It's usually added to help salt and other flavorings stick to the nuts. Planters has it in many of their varieties but several generic brands are gelatin-free.

Granola and Energy Bars: Any granola bar with chocolate will be clearly non-vegan because of the milk ingredients, but there are other things to watch for. They often have honey in them as a natural sweetener, and can have milk by-products like whey (more common in protein or energy bars than in granola bars).

Smoothies: Smoothies that you make yourself can easily be vegan but if you are buying a smoothie elsewhere, take care not to let dairy sneak in. Leaving out the milk or yogurt is obvious. You'll need to leave out the "protein powder" that can be added, because that is almost certainly going to contain whey.

Hidden Ingredients in Snacks

TIP

Read the labels! Most of non-vegan snack foods can be avoided by just choosing different flavors or brands. You won't have to totally give up on chips or peanuts.

Hidden Ingredients in Cosmetics

TIP

Avoid products tested on animals. It's not always obvious on the label, so do your research before you head to the store. And remember that **some cruelty-free brands may still contain non-vegan ingredients.**

Hidden Ingredients in Cosmetics

Like your clothing, your cosmetics or other body products may hide non-vegan things even though it's not all about food. You may not be thinking about your bathroom in your vegan quest, but there are actually a long list of ingredients you need to avoid.

Lanolin: Lanolin is a greasy secretion that is found on sheep's wool, and it's harvested by washing the fleece after shearing. It's sometimes listed as wool wax, wool alcohol, or wool grease in ingredient lists. It's used to make cosmetics smooth and creamy, and you can find it in many kinds of moisturizers and lip balms.

Beeswax: Most cosmetics that use beeswax are pretty open about it, so it's not even a hidden ingredient. Many natural products like soap, lip balm, lotions, and more can have beeswax.

Carmine: We mentioned this in the sweets section but it's found in various types of red cosmetics. Carmine or cochineal is made from crushed beetles. About 70,000 beetles are killed to make a pound of dye. There are lots of other ways to make things red, so there is no reason for companies to be using crushed bugs.

Gelatin: Again, another sneaky product we already saw in many foods. It's also found in a variety of moisturizing shampoos or face creams, and some types of nail products.

Tallow: This is one that's typically just found in soaps and sometimes deodorants. It's basically boiled-down animal fat, and it's sometimes listed as stearic acid, lard, fatty acid or benzoic acid. Glycerin is another form of fat that is very common in soaps and beauty products, but it is mostly from vegetable sources. Still, it would be worth looking out for before making a purchase.

Keratin: Nail products often have keratin as a strengthener. It comes from a ground up mixture of animal hair, feathers, hooves and other such animal parts.

Urea: As you might guess, this is derived from animal urine. Sounds like something to avoid even if you're not vegan. Though it sounds extreme, urea is actually very common in lotions, shampoos, deodorants and toothpaste. Some companies use synthetic urea, so contact manufacturers for details when you see this just to be sure.

Hidden Ingredients in Medical Products

Before we begin this section, let's be clear that we're not advocating that you should avoid vital or otherwise necessary medication because of its non-vegan status. That said, there are many examples of hidden animal ingredients in a variety of medical products.

Most Over-the-Counter Drugs: For the usual range of OTC drugs you may be buying in your household, you'll want to look out for gelatin, glycerin and colorants. Gelatin is an ingredient in any soft gel-cap style pills, though other forms of pills may use it as a binder. Glycerin is also part of some pills, though it can be a vegetable glycerin. Check with the manufacturer to be sure. Some may be labelled "vegetable glycerin." Colorings made from cochineal or carmine are tinted with crushed up beetle shells.

Certain supplements (often vitamin E) are made with fish oils. Some forms of vitamin D contain cholecalciferol, which is made from lanolin, a grease that comes from sheep's wool. Calcium supplements may have calcium carbonate, which may come from oyster shells, though there are several other sources.

In most of these cases, there are vegan options though they may not always be as available as the mainstream varieties. For the vitamins, it might just be easier to eat foods to boost your intake and avoid the supplements entirely.

Prescription Products: It was once the case where commercial insulin used by diabetics came from pigs or cows, that is no longer the case. Most insulin used today comes from genetically-modified bacteria. For many blood clot problems, heparin is used as an anticoagulant and it is currently still produced from slaughtered animals. Some forms of estrogen used in hormone replacement therapy come from pregnant mare urine, but there are other synthetic types available as well. Make sure you do your research.

Animal Testing: Unfortunately, almost all medicine has been tested on animals at some point in history, so you can't really avoid it completely. The biggest question is whether you are using products that currently are involved in animal testing.

Hidden Ingredients in Medical Products

TIP

If you are prescribed medicine that you feel contradicts your vegan lifestyle, **talk to your doctor about alternatives.**

The Clothes on Your Back

TIP

Don't assume that "faux" leather or fur is always environmentally friendly.

Do some research—plastic and other synthetic materials require petroleum & other non-renewable resources to create.

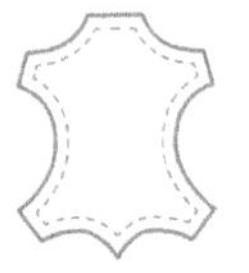

The Clothes on Your Back

We have mentioned that vegan concepts do go beyond your diet, and here is a prime example of what that means. Stop and consider the clothes you're wearing. Nothing may come to mind right away but there are a few common fabrics that are sourced from animals.

Silk: That's right, silk comes from insects and is a no-no for vegans. The fine fibers are spun by the Mulberry silkworm larvae to create their cocoons. Commercially, the larvae (caterpillars) are raised until they are mature enough to spin their cocoons. Then the cocoons are dunked in boiling water to kill the insect inside, and then the fibers are all unravelled to be used in fabric weaving. It will take the lives of 3,000 silkworms to produce 1 kg of silk fabric.

There are some varieties of silk (called ahimsa silk or peace silk) that are made from cocoons after the larvae have come out. The insects are not killed, and allowed to mature and live a full life as a moth after their work is done. But that doesn't sit well with most vegans because it is still exploitation and captivity.

Leather: Leather is usually the first material that comes to mind when you start thinking about animal products in clothing, so this shouldn't be a surprise. Though leather is a by-product of the meat industry, it still represents animals suffering in captivity.

Fur: This doesn't need much explanation. Many non-vegans even give up on fur because it's so often a purely decorative material that is harvested from animals that don't even get used for food. Animals that are trapped in the wild are usually not killed humanely, and those that are raised on farms don't fare much better. Many synthetic or faux furs on the market today are almost indistinguishable from the real thing.

Some Vegan Alternatives: Don't worry, there are several easy options for vegan materials. Cotton is the most obvious choice and its common enough that you don't have to look too hard. Various forms of synthetic fabrics like polyester is another option. Prefer to stick to something more natural? Look for bamboo, hemp or linen. For replacing leather, vinyl is a common alternative (check the tip on footwear for more on leather shoes).

The Shoes on Your Feet

We talked about clothing in the last tip, so it's only logical that we keep the theme going with a look at vegan shoes and footwear. This is where leather is going to be more of a problem, depending on your footwear choices. Dress shoes and boots are usually made with leather, though sneakers are made with other fabrics and are easier to veganize.

People often talk about the animal-based glues that are still used in the construction of shoes as well. Happily, that is actually an urban legend. Almost all shoes are made with synthetic glues these days. You can check with the manufacturer to be sure but it's highly unlikely to be an issue. So it's really just about the leather. Footwear can be made with canvas, rubber, cotton, vinyl and various forms of synthetic leather.

The reason shoes are trickier is that their materials are often hidden within the construction of the shoe, and a lot of different parts can go into some shoe designs. A seemingly all-canvas shoe may have a layer of leather under the insole or something like that. So sometimes it can be easier to find a footwear choice that is certified vegan rather than try to dissect the shoe looking for leather bits.

There are many vegan shoe outlets online if you can't find animal-free shoes locally. The large shoe shop Zappos.com carries a great selection (just search for vegan on their website). AlternativeOutfitters.com is completely vegan and they have footwear and accessory departments for men and women.

And if you're worried that you'll be walking around with wooden clogs on your feet, don't fret. Many vegan-friendly shoes are gorgeous and stylish.

TIP

Don't always go for new. Sometimes the most **Earth-friendly choice is to buy second hand** clothing or shoes, even if they're not made with vegan materials.

The Shoes on Your Feet

Be a Vegan Gardener

TIP

For vegetables you need to buy, the easiest way to verify what fertilizer or pesticides were used is to buy locally and ask a farmer.

Be a Vegan Gardener

Surprised that the world of gardening may have vegan concerns to it? Well, if you enjoy growing your food as a vegan, there are a few things you need to consider. Actually, the main concern is your plants' food: fertilizer.

In this age of organic thinking, we're all trying to avoid synthetic chemicals on our food, and that means more natural fertilizers are being used. And the most common one you buy at the gardening store is animal manure. If you think about the bigger picture, the only reason that this manure is around to be collected is because millions of animals are held in farms for food production. Buying manure helps to support the industry as a whole.

Unfortunately, there's more. Other common garden fertilizers include blood meal, bone meal, and fish meal. Obviously, these are all animal products you'll have to avoid. One last thing to watch out for is live insects. Granted, it's not the kind of hidden thing you might get tricked into buying because you don't know what's in the box. Even so, just be aware that garden stores often sell boxes of live ladybugs or praying mantises to be used as natural pest deterrents.

Your options? Stick with plant-based compost as your natural fertilizers. Can't find any in the stores? It's time you started composting right at home. Your garden will thank you. Just get a big bin, or build your own with some pallets or chicken wire (the Internet has loads of plans). Gather up all your kitchen scraps and yard waste, and dump it all in. Keep it damp, turn it over occasionally and you'll have excellent animal-free fertilizer by the next growing season.

What about your own insect issues? As a vegan, you won't want to kill off bugs in your garden just because it's convenient for your veggies. Your best bet is a strongly scented repellent that doesn't actually kill any bugs. It's a vegan choice that usually means less toxins in your garden. That makes it a win-win.

Level 3

Foods to Embrace

TIP

If you just want to try nutritional yeast, **buy it in a premade mix,** like vegan mac 'n' cheese.

Get to Know Nutritional Yeast

We've already told you that being vegan doesn't mean you have to eat a whole lot of strange foods. This is one very slight exception. Actually, nutritional yeast isn't all that strange, just unknown to most people. Time to get to know it.

To make it sound more appealing, some people in the healthy eating world have given it the nickname "nooch." It's kind of caught on, but don't be surprised if folks in health food stores have no idea what you're talking about if you say it.

As a side note, yeast of any variety is not an animal product. It's actually a fungus, and so it's perfectly fine for vegan diets. Nutritional yeast is not the same as brewer's yeast (for beer making) or baking yeast. It's a whole different product, and the others will not taste anything like nooch.

Why Eat It: There are basically 2 main reasons why you'll want to start eating nutritional yeast. The first is the taste. It has a very unique flavor that is a rough combination of nutty and cheesy. You can use it in many dishes as an ingredient or a topping. The other reason to embrace nooch is all the nutritional elements it has. This simple product adds zinc, folic acid and protein to your diet. That's great by itself, but the real star is B12. As a vitamin, B12 doesn't exist naturally in plants, so a diet that completely lacks animal products is going to be deficient. Just add some nutritional yeast to fill up that gap. Without B12, you are going to be at risk for anemia, so it's worth staying on top of it.

How Do You Use It: You can sprinkle some on top of pasta to replace Parmesan cheese, and it's a very tasty topping for freshly popped popcorn, scrambled eggs or hot baked potatoes. There are recipes for vegan cheese sauce that uses nutritional yeast too. A spoonful stirred into sauces will thicken them and add a richness to their flavor.

Where To Get It: You might have to do a little searching to find a good source of nooch. Many large health food or bulk food stores will carry it, as will some really good mainstream grocery stores. Bob's Red Mill has it, and the Red Star brand is also quite popular. If your local stores don't carry it, check online. Many shops will ship, if you don't mind paying postage costs.

So to boost your flavor as well as your vitamin intake, try a little sprinkling of nutritional yeast when you cook.

Cooking with Vegan Proteins

Once you eliminate meat, eggs and dairy from your diet, you'll need to take care to get enough protein. It's not as hard as critics would make you think. Many beans, nuts and seeds are packed with protein for example. But if you are looking to get a more substantial serving of protein, and possibly replace the texture and sensation of eating meat, there are some vegan options.

Tofu: This is the most common meat-replacing protein, and it's really easy to find in just about any supermarket. It comes in various textures from really soft or silken, to very firm. Tofu pretty much has no taste whatsoever on its own and will blend in perfectly with any recipe. The silken tofu is fabulous to thicken up smoothies or other desserts, and the firm can be sliced into a stir fry recipe to replace the meat (just marinate first if you can). Worried about GMO soy? Go for organic tofu then. Regardless of the specifics, you'll get around 10g of protein per cup.

Tempeh: A close relative to tofu is tempeh, though you may not have heard of it. Made with soy using a different fermentation process, tempeh is much firmer and has a nutty or mushroom-y flavor of its own. In terms of texture, this is a great one to replace meat though the flavor is more likely to stand out than it would with tofu. Check the refrigerated section at the health food store. There is a lot more protein in tempeh, with more than 30g in a cup.

Seitan: Of all the ingredients in this tip, seitan is the most meat-like in texture. It's actually quite a bit like bland white chicken meat. It's also a great choice if you don't want to load your diet up with a whole bunch of extra soy. Seitan is made with wheat gluten, so you will have to avoid it if you have Celiac's disease or other gluten intolerance. Unlike the other two, you can actually make seitan fairly easily yourself with flour (or straight wheat gluten, sometimes found in bulk food stores). For protein, it's a winner. One cup of seitan has 75g of it.

Quinoa: This isn't really a "meat substitute" in texture, it's much more like a grain. It fits with this page because it has a significant amount of protein in it, so you can whip up your favorite rice dishes and boost the protein by using quinoa instead. You'll get 24g of protein in every cup, which makes it even better than tofu.

Considering the average woman needs 46g of protein per day, and the average man should have 56g, you can see that it's not hard at all for a vegan to stay healthy without meat.

Cooking with Vegan Proteins

TIP

Here are a few other examples of foods
you can get your precious protein from **(all for cup-sized portions)**

black beans - 42g / **chickpeas - 39g** / **lentils - 50g**

Cooking More Beans

TIP

Beans You Should Try

black beans
kidney beans
chickpeas
(or garbanzo beans)
navy beans
lentils **(red, green or whatever color)**
adzuki beans
fava beans
white beans

Cooking More Beans

We finished our last tip off with a nod to the great protein you can find in various beans, so why not continue the discussion? Here is another tip that involves what you can eat rather than what you can't.

Beans and other legumes can be the new nutritional foundation for your meals once you give up on meat. It can be hard to give up that heavy texture once you've lived your whole life eating meat. Though delicious, the crisp feel of vegetables isn't always enough to satisfy when you're hungry. That's where beans come in.

Cooking Bean: Using lots of canned beans is the easiest way to get into legumes, as they are already cooked and ready to go. Just add them to soups, stews, chili, or rice dishes. Simmer until hot, and that's it. You can save money and keep your dishes a little healthier if you use dry beans, though. Canned beans usually have salt added, and even sugar on occasion.

Cooking with dry beans means planning ahead (not always easy for the busy vegan). You'll need to soak and simmer for hours, possibly even overnight. Lentils are really the only ones you can cook right away without the soaking.

Bean Nutrition: Aside from the filling satisfaction you get from a bean-laden meal, there are several strictly nutritional reasons why you should eat more legumes. They're loaded with protein. You can get your daily requirement with just 1 cup of cooked lentils. Add in huge amounts of fiber, iron, folate, calcium, magnesium, and vitamins A, C, and several of the Bs. We all know that green veggies are healthy, but it's often a shock to hear just how much goodness is in the humble bean.

Eating More Fruit

With its naturally sweet and juicy texture, fruit tends to be left to desserts and treats in the typical diet. This section is about sweet fruit, not things like pumpkins or tomatoes.

The Nutrition of Fruit: The list of vitamins you can find in fruit is almost endless. Vitamins C and A are two of the biggest, but you'll also get lots of fiber, potassium, folic acid, and many kinds of antioxidants. Various forms of B vitamins are also found in lots of fruits, though in somewhat smaller amounts.

Compared to vegetables, there is a lot of sugar in fruit, so you do have to watch to make sure that you don't overdo it. Seeing as vegans don't eat the usual refined white sugar, though, it can be a nice way to get some sweetness into your dishes.

If you're looking for ways to boost your iron intake, try to snack on raisins, dried apricots, or dried peaches. They will provide around 10% of your iron requirement in just a big handful.

Avocados: You can't talk about eating fruit without giving a little attention to the unique avocado. It holds a special place in a vegan heart because of its nutritional makeup. Unlike most other fruits, you'll get healthy fats, protein, and calcium from an avocado, not to mention a creamy texture that can be used in so many "non-fruit" dishes. You can thicken many different things like frosting, sauces, smoothies or even pudding without having to use dairy. Give it a try.

Not Just for Dessert: The trick to adding more fruit to your overall diet is to get over the idea that fruit is only for the end of the meal. Add bits of fruit to salads, or even to a main dish. Stewed or broiled fruits can add excellent flavor to an otherwise savory dish as their sweetness tends to diminish when cooked. Any kind of baked good can be jazzed up with dry or fresh fruit.

Fruit is so healthy, there are some people who have limited their diets to only include fruit. The term is fruitarian and it's considered an extreme form of veganism. Some people choose to eat only fruit on ethical grounds because it can be harvested without killing the plant. It's an interesting viewpoint when you stop and think about how to respect life.

TIP

Fruit makes a great addition to salsas, salads, and even pizzas!

Eating More Nuts and Seeds

TIP

Remember that you need healthy fats for a balanced diet. Nuts are a great source of it!

Eating More Nuts & Seeds

Nuts and seeds can bring a lot to the table for a vegan. They're not just for snacking anymore.

Nutrition in Nuts: Like with beans, you can get a lot of the nutrition normally associated with animal products from nuts and seeds. You'll find great doses of protein, fat, and fiber in many nuts. Add in large amounts of calcium, magnesium, and iron and your body will never even miss the meat.

One cup of almonds will give you nearly half your daily protein requirement. You'll also get around 40% of your daily fiber and 25% of your calcium. Not bad at all.

Cooking with Nuts: Nuts aren't always a big addition to most cooked dishes, but they work nicely as salad toppers or as ingredients to baked goods. Chopped nuts can be added to just about any kind of muffin or cake for some nutritional boosting. And if you don't really want to cook with nuts, you can just stick with a handful of nuts or a lovely trail mix.

Seeds are a little easier to work into your cooking. Flax, sesame and poppy seeds are a few that can go into a multitude of recipes. They're not quite the same powerhouses as nuts but will still add flavor and fiber to your diet. Pumpkin seeds and sunflower seeds are amazing in baked goods too.

Nut Butters: Another way to add more nuts to your diet is with tasty nut butters. We all know peanut butter, but you can change up the taste with cashew or almond butter too. You do have to watch for sugar content in some cheaper brands of peanut butter that might not be vegan. Good-quality peanut butter (or any other nut butter) shouldn't have sugar anyway. Smoothed on toast, bagels, or in PB&J are just a few of your options.

Replace Your Dairy: One of the interesting things about nuts is that you can use many of them to replace your dairy products, which can be a bit of a surprise. With a food processor and a supply of nuts such as cashews or almonds, you can create a whole range of vegan "dairy" products that are healthy and cheaper than store-bought ones. Soak whole cashews for a few hours then blend with a little more water, and you can have a delicious vegan milk replacement. A similar approach can be used for almond milk too.

Cooking More Leafy Greens

Greens aren't just for salads! Of course, salads are delightful and delicious but you can add lots of variety to your entrees with cooked greens too.

Leafy Nutrition: You can add a lot of vitamins and minerals to your meals with any combination of greens, though iceberg lettuce is pretty useless. Depending on the type, you can find significant amounts of folic acid, calcium, iron and vitamins A, C and K in many dark leafy greens. You'll also get fiber and many kinds of antioxidants.

Branch Out: If the only greens you're familiar with are lettuce and spinach, you're in for a treat. There are loads more varieties that will bring new flavors and new textures to your food. Be on the look-out for:

- **Kale**
- **Swiss Chard**
- **Arugula**
- **Collards**
- **Mustard greens**
- **Turnip greens**
- **Romaine lettuce**
- **Dandelion**
- **Cress**

Get Cooking: Eating these veggies raw will give you the biggest nutrient boost but cold salads can be tiring after a while. To help fill up your main dishes now that you've taken out the meat, add more cooked greens instead. The usual way to cook leafy greens is a light steaming. That softens everything up nicely without cooking your leaves down to mush.

You chop up your greens (including stems for things like kale or chard) and get an inch of water boiling in a pot. Either use a steamer basket or wire sieve in the pot to suspend your greens over the boiling water. Cover and let it steam for a few minutes, until your greens are just the texture you like. Kale can take up to 8 minutes but spinach may be ready in just 3 minutes. Change up the flavors with additions of soy sauce, garlic, nutritional yeast, chili sauce, or any other seasoning that tickles your fancy.

You can also add lots of greens to other dishes without having to cook them separately at all. Chopped greens can be added to all sorts of foods, like chili, lasagna, soups, stir fry and more.

Cooking More Leafy Greens

TIP

Get some veggies in your smoothies! A few tablespoons of chopped raw greens are practically invisible in a fruit-filled smoothie & add some extra fiber and nutrients with every sip.

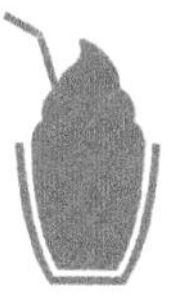

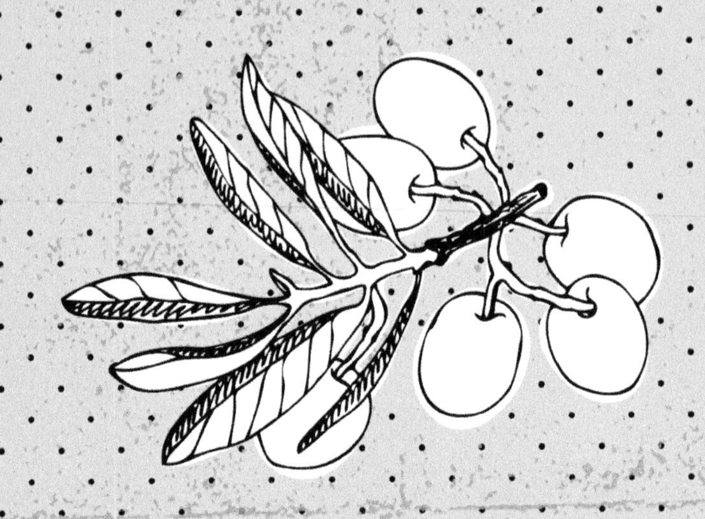

Vegan Probiotics

TIP

You can also **get vegan-friendly supplements** to bring more probiotics into your life.

Vegan Probiotics

If you are up-to-date with current healthy eating trends, then you've probably heard of probiotics. These are a big mix of "good" bacteria and other single-celled organisms that are super helpful in helping your digestive system stay healthy and efficient. Do you have to give up probiotic products if you are vegan?

That's actually a trickier question than you might think. The organisms that are grouped as "probiotics" are a mix of yeasts and bacteria, and none are considered to be animals. So, the little critters themselves are fine for vegans. The problem comes with your options for getting them. The standard products on the shelves today for probiotics are almost all dairy-based, particularly yogurt and kefir (a fermented milk beverage). Those of us not eating dairy, where do we go?

The yogurt companies may not want you to know this, but you can get those probiotics in all kinds of plant-based foods. Some of them may already in your diet already. You need to look for anything fermented or kept in brine. That's where the good bacteria like to hang out. Foods such as:

- **Miso soup**
- **Tempeh**
- **Olives**
- **Soy milk**
- **Kombucha - a fermented tea beverage**
- **Sauerkraut**
- **Kimchi - a Korean dish similar to sauerkraut**

One little aside about lactobacillus. The name of this one bacteria trips up vegans because the "lacto" gives the impression that it is a dairy milk product. Though it is commonly found in milk (hence the name) that is not the only source for it. Lactobacillus lives all over the place, and seeing it as an ingredient or part of a supplement doesn't necessarily mean the product is non-vegan. Still, look for the vegan logo to make sure it's not dairy-sourced.

Getting Calcium & Iron

Certain nutrients are always associated with a meat-based diet: protein, calcium and iron. We've already talked quite a bit about protein a few pages ago, so now it's time to tackle the other two.

Calcium: Without milk, where will you get your calcium? Well, lots of places actually. Adults need approximately 1,000mg of calcium per day though teens and seniors should have closer to 1,200mg. To keep your bones and teeth healthy, you certainly be aware of your calcium intake and make sure to eat some of these foods often:

- **1 cup soy with added calcium - 300mg (same as cow's milk)**
- **1/2 cup of tofu with calcium sulfate - 800mg**
- **1 cup orange juice with calcium - 260mg**
- **1 cup raw kale - 95mg**
- **1/2 cup cooked turnip greens - 100mg**
- **1/2 cup black-eyed peas - 185 mg**
- **1 tablespoon blackstrap molasses - 170mg**
- **8 dried figs - 105mg**

The easiest way to boost up your calcium is to go for anything fortified with added calcium, like the soy milk (other non-dairy milks are good too), tofu, and orange juice. Add in a few of the other veggies and your body will never miss the cow's milk.

Iron: This is one nutrient we haven't talked about too much yet. Iron is used by your body for a few things, but the most important is to maintain your red blood cells that carry oxygen. Without iron, you can run the risk of anemia, though the recommended amounts are fairly low. Adult men need just 8mg and adult women need 18mg. The typical source for iron is red meat, so where do vegans go for their iron?

- **1 cup cooked soybeans - 8.8mg**
- **1 cup spinach - 6.4mg**
- **4 oz tofu - 6.4mg**
- **1 cup cooked lentils - 6.6mg**
- **1 cup swiss chard - 4mg**
- **1 cup black-eye peas - 4.3mg**

Getting Calcium and Iron

TIP

The bottom line is that you need iron *AND* calcium.

If you find you can't manage with the plant-based sources, go for a supplement.

Even Junk Food!

TIP

These products can change their recipes or ingredient lists at any time.
You should still double-check the contents of any food.

Even Junk Food!

While you may not really want to "embrace" junk food, this is still a section about foods that do fit within a vegan lifestyle. And let's face it, indulging in junk now and again can be a treat.

Believe it or not, there are actually several kinds of mainstream junk food that are vegan. You don't necessarily have to search out the intentionally vegan brand lines. Here are a few things you can snack on:

- **Oreo cookies**
- **Ritz crackers**
- **Ruffles and Lays potato chips (the plain ones)**
- **Swedish fish candy**
- **Some flavors of unfrosted Pop-Tarts**
- **Fritos corn chips (original flavor)**
- **Several flavors of Nature Valley crunch granola bars**
- **Wheat thin crackers (original flavor)**
- **Doritos (just the spicy sweet chili ones)**
- **Krispy Kreme fruit pies**
- **Manwich sauce**
- **Cracker Jacks**
- **Some flavors of Ghiradelli hot chocolate**
- **Red Bull**
- **Pringles (original)**

As you can see, being vegan doesn't necessarily equate to being healthy. At least you won't have any guilt about animal welfare when you're snacking on these. Just guilt about your waistline.

And there are many great vegan brands of snack foods you can try if you'd rather not support companies that sell foods with animal ingredients. Check out the food section at the Pangea online shop (veganstore.com) and you'll find plenty of tasty treats from vegan producers. And for the eco-minded vegan, keep in mind the production processes that big companies use to make those junk treats.

Level 4

The Vegan Lifestyle

Dealing with Non-Vegans

TIP

Be respectful of other people's choices and share your thoughts on veganism in a more conversational way, especially when answering questions.

Dealing with Non-Vegans

There's a well-known joke about vegans that goes something like this:

How do you spot a vegan at a party?
Don't worry, they'll tell you.

It's not particularly hilarious but it does nicely illustrate the image that vegans have to the general public. Vegans are known to be pretty vocal about their beliefs, and not in a good way. We're considered to be pushy and preachy. It's not a great image, and it's time we tried to change it.

One school of thought is that the best way to educate others about animal cruelty in the food industry is to tell them, to make a point of showing off your vegan lifestyle as often as you can. Well, it can help let people know but if you scare them off after 30 seconds, it's not really going to help. If you're at a BBQ, enjoying your black-bean burger, no one is going to want to sit with you if you keep telling everyone that their beef burgers are contributing to the horror of systemic animal torture.

If you're invited to a party or gathering, don't get on a high horse and insist the host provide you with a customized vegan menu. The other extreme would be to show up, then make a show of not eating anything except a plain salad. Be a good guest and deal with it. Ask what's being served, and see if it's ok to bring a few items of your own. Few people would mind if you brought a vegan burger or hot dog.

Now that's not to say you have to hide your thoughts completely, just be reasonable and polite when voicing them. If someone asks why you don't eat meat, you don't have to lie about it. Say you don't like to support the way animals are treated and see if the conversation goes anywhere. Describing the slaughtering process in graphic detail is going too far, unless you're really getting into a debate with someone. And if you are debating, don't let it devolve into fighting.

You will sometimes have to deal with non-vegan gifts. Nothing alienates friends and family faster than a refused gift, so tread lightly here. Before known gift-giving occasions (like Christmas), make it clear what isn't welcome in your home. Better yet, focus on what is wanted instead. Ask for books, gift cards, jewelry, kitchen gadgets, home decor items, donations to charity or anything else that frees up the gift-giver from dealing with animal by-products. If people do give you non-vegan items, be as gracious as you can.

Can Vegans Keep Pets?

This is a bit of a tricky subject and you'll probably find that not all vegans have the same answers for you. Just consider this tip something to think about and you'll have to come to your own conclusions on how to tackle the issue of pets in a vegan home.

If you're a vegan because you like the personal health benefits of a strictly plant-based diet, then the idea of keeping a pet probably won't be much of a concern for you. The issues come up if you are vegan because of animal treatment issues. Does the belief in animal welfare, and that animals shouldn't be exploited by humans extend to your pampered cat or beloved dog? Some would say yes, and some say it's not the same thing at all.

The next question is all about your pet's diet. Even if you consider having a pet to be an acceptable vegan practice, what about the animal products in their food? This is where most people start to get seriously hung up. Your cat may be living a comfortable life, but animals that provide the chicken, beef and pork for your cat food probably aren't.

Some people make the situation work by feeding their pets a vegan diet but it's questionable about how healthy that is for them. Humans are omnivores by nature and are designed to handle lots of plant-based food. Cats and dogs are carnivores and simply aren't as well adapted to surviving on plant proteins. If you do want to go that route, talk to your vet about your options.

So you may want to consider a herbivore as a pet to get around this problem. A rabbit, various small rodents, or even fish might fit your lifestyle better. Many vegans choose not to keep pets at all, which is a perfectly reasonable decision given their beliefs.

Of course, the situation changes if you already have a cat or dog. Kicking them out of your home is not really an option so you may have to deal with buying them animal-based foods for the duration of their lives.

Can Vegans Keep Pets?

TIP

Look into eco-friendly ways to feed your carnivorous pet. Do your research on how high-quality pet food is made or consider homemade food—but always consult your vet.

Taking It on the Road

TIP

Don't blow your top because a waiter doesn't immediately know if the salad dressing has egg in it. **Always be gracious about your food concerns.**

Taking It on the Road

Eating out near your home is bad enough when you're vegan, but once you head out into less familiar territory, it's going to get harder. Here are some tips on being vegan when traveling.

Pack Snacks: Snacks or any other portable food that you can have on hand for those times you just can't deal with finding a vegan meal. Sit-down restaurants can usually accommodate your diet to some degree, but it's the quick snacks that can really get you. Dried fruit, nuts, vegan crackers or granola bars are just a few travel options. Ideally, your snack packs shouldn't need a fridge if you can avoid it. If you're planning on bringing perishables (maybe some vegan cheese or rice milk?), bring a cooler with ice packs.

Research First: Before you take a step out the door, find out what you can about any restaurants or food shops at your destination. Check websites, brochures or make some phone calls to see what kinds of dishes are on the menu. Once you know which places offer at least a few vegan-friendly meals, it will make life easier once you're there.

This applies doubly for fast-food joints. You're not likely going to have more choices than a salad or two, or maybe french fries. Depending on the route you're travelling, you may find that fast-food chains are your best bet (perhaps only bet) in some cases. These places are also good for your budget when you don't want to spend a lot with each meal. The next section has some more tips on dealing with restaurants in general.

Get Cooking: When you're moving from place to place, this probably won't help much but once you've settled at your destination, get a room with a kitchenette. Trying to rely on unfamiliar restaurants can be stressful and not always that successful, depending on the city you're in. No one wants to do all the cooking when they are on holiday, but it does mean you can control your food options, and you'll likely save a little money while you're at it.

Plan out your recipes beforehand, and have some ingredient lists packed with you. Check with the hotel to see if any kitchen equipment is included. You may need to pack a few things to make this work.

Be Patient: It can be tough being away from home, and having to sniff out decent places to eat can wear on anyone's patience after a few days. Plan ahead to reduce stress and wasted time.

Eating Out

Eating out at a restaurant can be a challenge to anyone on a unique diet, vegan or otherwise. It can be hard to manage the situation when you have certain requirements yet someone else is in control.

Plan Ahead: This is the biggest part of a successful dinner out, even though it might spoil a little of your evening spontaneity. Check out websites, read menus, make phone calls before you get there to figure out what may or may not be vegan friendly. Trying to figure out your choices while sitting at the table, probably while hungry and impatient is never a good idea. Once you establish what dishes you like at your local restaurants, you can just stick with them without having to go through this every time.

Pick a Reasonable Place: Trying to find vegan food at a steak house or a BBQ joint? Why do that to yourself? Some restaurants are simply going to be meat-heavy and you're in for an uphill battle trying to deal with it. Unless you're fine with a garden salad, don't bother. Instead opt for places that have wider menus, particularly those that embrace a healthier attitude toward their food. Also try looking into cuisines that are naturally more plant-based, such as Chinese or Indian.

Munchyy.com has a great list of all the vegan items on many chain restaurant menus (http://munchyy.com/43-vegan-chain-restaurant-menus-every-vegan-needs-know/), though some are limited to just drinks and condiments. It's still a good spot to check when it comes to fast food options.

Dishes to Focus on: Hoping that a chef can somehow make a stuffed chicken breast without the chicken is a little silly, so get to know the best meatless dishes you often find on a menu:

- **Vegetable stir fries**
- **Salads**
- **Pasta dishes with non-cream sauces**
- **Wraps or sandwiches, hold the meat**
- **Fries as long as they are not fried in the same oil as meat items**
- **Pizza, opt for vegetarian without cheese**
- **Chips and salsa appetizers**
- **Vegetable side dishes**

These types of food are often vegan, or can be made vegan with only a few small tweaks.

Eating Out

TIP

Invite friends over for dinner, rather than going out.
At home, you know you can serve healthy vegan food to your liking
—and you may save a few bucks!

Raising Vegan Kids

TIP

Vegan parenting isn't dangerous, but always involve your doctor.

Raising more children to make animal-friendly vegan choices will build eco-friendly habits for the future.

Raising Vegan Kids

As plant-based lifestyles become more and more popular these days, a new generation of vegans is being born. Some thoughts on raising new little vegans:

Know Childhood Nutrition: Even the most experienced vegan parent should brush up on the nutritional needs of young children. Protein, vitamins, and minerals are crucial and often needed at different levels than for adults. Though many publications and professionals will be critical of a vegan diet for children, the reality is that it can be perfectly healthy and beneficial.

Be Prepared for Commentary: People already have pretty strong opinions about vegan living, and once they hear that you're raising your children that way, you're going to hear even more. Don't get your back up, and either let the comments go or have some good responses ready. Having some good facts about nutrition will help diffuse any arguments.

Be Gentle About It: With your kids, that is. Telling them that your family doesn't eat anything from animals because of the way the poor creatures are tortured is just going to traumatize your children. Keep your explanations simple and age-appropriate. Perhaps focus on how healthy and yummy the foods are in your house.

Keep It Kid-Friendly: Tasty treats that make kids happy are easy enough to manage in a vegan household. You can still have cupcakes, hot dogs, fries, chips, and candy without any animal ingredients. Sure, these aren't the healthiest foods out there but sometimes that's exactly what a kid needs.

Let the School Know: With so many food allergies out there these days, schools are a lot less likely to allow food in the classroom without parental knowledge. Even so, let your school know that your kids are vegan and that they shouldn't have any non-approved treats. So your children aren't left out, stay on top of any school parties and offer to send along some vegan goodies.

Let Family Know: This one is pretty obvious so you're not having to deal with inappropriate food choices from family members (you may have to put your foot down if they won't respect your choices). Just don't forget to tell them that you also won't want any clothing made from wool or leather either. Many folks won't be aware of the non-food vegan considerations.

Should You Date a Non-Vegan?

This isn't an easy question to answer, and not everyone would agree to a universal solution anyway. If you're struggling with this, here are a few things to consider. But ultimately, you'll have to decide for yourself.

The dating itself can be manageable, provided your potential partner doesn't try to take you to the zoo or the local steakhouse. The bigger question goes beyond a few dates. If the relationship gets more serious, then what? Putting up with the habits of a non-vegan may be tolerable for an evening or two, but could you honestly see yourself living with someone who still eats meat? Besides the complexity of managing a kitchen with two such diverse diets, it can be emotionally difficult to be with someone who supports the abuse of animals in the food industry.

Maybe they can change? Well, it's not fair to say this can never happen. Some people may be intrigued by your vegan lifestyle and open to taking it on for themselves. On the other hand, someone who is really not interested is unlikely to change their diet so drastically just to make you happy. They have to want it for themselves.

Some vegans focus more on the health reasons for their plant-based diets, and that can make communing with a carnivore a little easier because their habits don't impact your body. If you are like most other vegans with a strong concern for animal welfare, it might not go as smoothly.

If you're not opposed to it, try to compromise by asking your partner to eat less meat and animal products. It still has a great impact on the environment and it gives you a chance to introduce them to more vegan dishes.

Think of it like dating someone of a different religion. If you're constantly worried about someone else's soul, you're never going to be happy.

Should You Date a Non-Vegan?

TIP

It's very difficult to just "live and let live" when you're extremely passionate about a particular issue, and that's okay. **You get to decide what your priorities are.**

Being Vegan on a Date

TIP

Dating locations shouldn't be a problem, but you may want TO avoid any local zoos or aquariums.

Being Vegan on a Date

We just talked about choosing whether or not to date a non-vegan, and now it's more about the nitty-gritty of being vegan while on a date. Even if you are dating another vegan, it can be tricky to navigate all the various date-worthy activities without causing undue stress to either of you. Here are some options.

Movies: The classic date activity, and one that should fit nicely with any vegans. There is actually a nice selection of goodies you can enjoy at the snack bar, including the classic box of movie popcorn (provided they don't use real butter). Many candies are fine but you should plan out your favorites before you go. The chapter here on hidden ingredients in sweets can help point you in the right direction. Milk chocolate and most brands of gummy candy are probably out.

Dinner Out: You'd best speak up if you're heading to a restaurant for your date. If your date suggests heading to the local BBQ shack, you're going to have problems. Make it clear that you need a place that has vegan-friendly options. You also need to face the fact that your date is likely to order something with animal products in it, unless they're vegan too. If you really feel that is going to make you uncomfortable, you might want to reconsider the date in the first place.

Going Out for Drinks: Hanging out at a bar is a fun and casual evening, and it can be a perfectly vegan one as long as you watch what you order. We discussed this in more detail in another chapter and you can research which drinks are off-limits. Just know what you can order before you head out the door, and your evening should go smoothly. Wine is often non-vegan but you can choose from a number of beers and other cocktails.

Telling Your Date: Depending on what your plans are, you may or may not want to announce your veganism right away. You're not defined by such a label, and you might not want to make that the most important thing about yourself to reveal. That's really up to you. At some point though, make your lifestyle choice clear so that future plans can be appropriate. If you think it's going to be a real problem for the other person, then perhaps you need to find another person to date.

Being Vegan on the Holidays

Specifically, all those family holidays that revolve around food. Unfortunately, that tends to mean most holidays.

Regardless of the specific holiday, it's always best to be a respectful guest when visiting family at the holidays. The broader tip on dealing with non-vegans really applies here, so make sure you've read that too.

Christmas & Thanksgiving: These two holidays are very similar when it comes to food and family traditions, so you can really take a few suggestions that would apply in either case.

Talk to the host or hostess ahead of time. Rather than try to convince them that the family would certainly love to try a vegan tofurkey this year, keep your requests simpler. Find out what's being served, and ask for small adjustments. Perhaps the mashed potatoes can be made without milk or butter, or the salad can be served with vegan dressing.

Help out by bringing a casserole or hearty side dish that would work well as your own meal, even if nothing else on the table is edible for you.

Easter: Though it's a huge family holiday, it's not quite as food oriented as the first two. The same basic tips above apply for a dinner get-together at Easter too.

Then you have all the chocolate. Because of the various milk ingredients, many forms of chocolate are not vegan. There are a few brands of vegan chocolate (Dagoba and Green & Black for starters), and some of the darker varieties of Lindt chocolate are also vegan. The easiest work-around is to just ask for non-chocolate sweets from anyone would give you Easter treats. Same goes for anything you might give to others.

Birthdays: For your own birthday, you should feel a lot freer to make "demands" about what sorts of food you want served. Any unsuitable gifts you receive can always be regifted or donated without causing a fuss.

One last tip that is somewhat unique to family holiday meals: don't rise to the bait. Friends and other guests are generally too polite to really get in your face about being vegan. Family members aren't always so nice. When that loud-mouth uncle waves a drumstick under your nose and comments about how delicious it is, do your best to let it go. If you really know people are going to be judgemental, maybe you should think twice about going at all.

Being Vegan on the Holidays

TIP

For someone's birthday, offer to bring a platter of vegan cupcakes to a casual party. You'll have something to snack on AND can show off a great vegan recipe.

Animal Entertainment

TIP

Remember the plight of animals outside of the kitchen.

We can easily forget the many others that are just as captive.

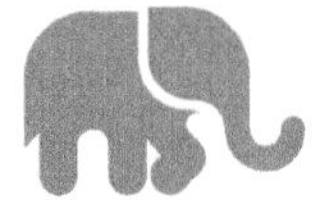

Animal Entertainment

This is one of the few areas outside the home where your vegan ideals will come into play. Animal captivity and poor treatment goes beyond the food and clothing industries, to include all forms of entertainment such as aquariums, zoos and circuses.

Even non-vegans often protest the treatment of animals in the circus, particularly because the venues are strictly for entertainment and the constant moving of the animals is so stressful for them. Forcing animals into tiny enclosures and then to perform tricks in a loud unfamiliar environment is cruel no matter how you look at it.

As awareness of circus abuse has grown in the past years among the general public. Recently, the Ringling Brothers circus has decided to retire its elephant acts altogether. Unfortunately, the other animal acts are staying in the lineup, though with further public pressure, that could change. Ironically, one of the largest circus acts in the world, Cirque du Soleil, has no animal acts whatsoever. It's certainly possible to see an amazing show without putting animals through misery.

Zoos are a little less cut and dry, because they do serve a scientific purpose to some degree. Preservation, breeding, and public awareness of endangered animals are all part of many zoos' missions. But many smaller zoos are strictly around to entertain the public, and animals are kept in small or otherwise inappropriate enclosures throughout their lives. Young are separated from parents, and frequently moved from facility to facility as "inventory" needs shift each year. Rather than maintaining zoos, money would be better spent protecting animals in their natural environments in the first place.

Aquariums are even worse than zoos. A zoo may somewhat replicate an animal's natural habitat, but there is no way a tank can even come close to mimicking the ocean. Whales and dolphins are extremely social and intelligent animals, and keeping them in these tiny tanks is beyond cruel. Many of the animals suffer physically and psychologically from captivity.

Groups to Support

If you want to make a difference outside of your personal habits, you may want to get involved with animal welfare organizations. Whether it's just through membership dues, donations, signing petitions, or volunteering, these are the kinds of groups you can support to help make an impact on animal welfare in a bigger way.

The Vegan Society: They are the group behind the international vegan symbol on foods. Though they are a UK-based group, they advocate for the vegan lifestyle and animal welfare around the world. You can donate as a one-time gift or join the society to get their newsletters.

PETA: Though this might seem like a very obvious choice at first, the works of PETA can be contentious among vegans and non-vegans. They're well-known for shock tactics like throwing paint on fur-wearing celebrities. Even if you don't want to support them financially, they are a great resource for keeping up with the latest in animal welfare and political activism, like petitions.

Your Local SPCA: The Society for the Prevention of Cruelty to Animals isn't strictly a vegan cause, but it does do a lot of good towards addressing poor treatment of animals (usually pets, but not always). They are run by volunteers so they would be happy to receive money as well as a commitment of time to help keep their shelters open. If you don't have a local group, you can also direct donations to the larger ASPCA (the nation-wide American SPCA).

Radical Activists: There are a few other animal rights groups that you might hear about, and you should know the details on some of them before you get involved. Usually labelled as eco-terrorists, organizations like Earth First! and the Animal Liberation Front (ALF) are known for their drastic tactics, usually involving illegal activities such as destruction of logging or mining sites, breaking into labs to free animals and other such practices. Some feel that serious efforts are needed to save or protect animals, but these methods tend to alienate "average" vegans and make our cause difficult to justify.

Other possible groups that might catch your interest include **Mercy for Animals, Farm Animal Rights Movement (FARM), World Society for the Protection of Animals (WSPA**) and the **Movement for Compassionate Living.**

Groups to Support

TIP

The kinds of groups you support depends on your own personal beliefs. **Not all vegans have identical mindsets,** so expect some differing opinions. Lots of them.

Level 5

Recipes

Spicy Black Bean Burgers

TIP

Sometimes your vegan burgers can get a bit crumbly.

To replace eggs, the usual binder, use oats, bread crumbs, or ground flaxseeds.

Spicy Black Bean Burgers

Soy burgers can be a great option when you're first going vegan but you can also make a healthy and home-made version yourself? Who needs beef when you have beans. Not only are these burgers packed with nutrition, they are baked rather than fried.

- **1 small onion, diced**
- **4 cloves of garlic, minced**
- **1 bell pepper (red or green), diced**
- **2 cans of black beans, rinsed and drained**
- **1 tsp coriander**
- **1 tsp cumin**
- **1 tsp chili powder**
- **1/4 cup rice flour**

Get the oven heating up to 350F before you start.

In a pan, saute the onion until it starts to get soft and then the garlic and chopped pepper goes in. Keep cooking for around 5 more minutes. Now mix in the black beans and spices. Continue to cook for another 3 to 4 minutes, then pour the whole batch into a big bowl. Use a fork or other masher to break the beans up. How chunky you leave them is up to you. Stir in the rice flour.

Form into 6 patties and bake for 20 minutes on each side. Serve up with condiments on a vegan bun.

Tofu Pad Thai

Pad thai is a zesty noodle dish that is often served with tofu, making it ideal for a vegan diet. If you're not familiar with cooking tofu, this makes a tasty introduction.

- **1 pkg. rice noodles (10 oz.)**
- **2 tbsp. olive oil**
- **2 cloves garlic, minced**
- **6 oz. extra firm tofu, cubed**
- **5 tbsp. soy sauce**
- **2 tbsp. peanut butter**
- **3 tbsp. lime juice**
- **3 tbsp. raw sugar**
- **1 cup water**
- **Chopped green onions (optional)**
- **Chopped peanuts (optional)**

Cook the rice noodles in simmering water for about 10 minutes, or until tender. Set that aside and heat up a large skillet with the oil. Sauté the garlic briefly, then toss in the tofu pieces. Continue cooking with a little soy sauce until the tofu is nicely browned. Be careful not to break up your tofu chunks too much.

Meanwhile, in a mixing bowl, stir rest of the soy sauce, peanut butter, lime juice, sugar and water together to make your sauce. If you want a little more heat, add a dash or two of your favorite hot sauce. Transfer your cooked noodles to the skillet with the tofu, and mix with the peanut sauce. Keep cooking until the noodles are heated through. Serve topped with a sprinkling of onion and peanuts.

Tofu Pad Thai

TIP

Pad Thai lends itself to variety, so don't make the same dish every time. There are plenty of recipes for sweeter or spicier versions.

Mock Tuna Salad

TIP

Spice up your salad with dried or fresh fruit for a sweeter taste. AND for an alternative to vegan mayo, hummus is a good option.

Mock Tuna Salad

What would a summer picnic be without a tuna sandwich? Well, even vegans can enjoy this tasty lunch by making a similar tuna-less spread with chickpeas instead. You won't even miss the tuna.

- **1 can of chickpeas (or garbanzo beans)**
- **1/4 cup vegan mayonnaise**
- **2 tsp. celery seeds**
- **1 tbsp. mustard**
- **2 tbsp. rice vinegar**
- **1/3 cup celery, diced**
- **2 dill pickles, chopped (optional)**
- **2 green onions, chopped (optional)**

Drain your can of beans and give them a good rinse. Use a good processor to mash or roughly chop the chickpeas. Now just stir in everything else and mix until it's combined to your liking.

Now serve on buns, bread, or just with a salad.

Tofu Scramble

If you like a hearty or savory breakfast, it can be hard to give up on your favorite scrambled eggs. Firm tofu is a lot like cooked egg, and a good pan of scrambled tofu will start off your day just right. There are lots of different ways to make this dish, but here is a good simple recipe to try. You can experiment with your own mixes of vegetable and spices once you get the hang of it.

- **1 pound extra firm tofu**
- **2 tbsp. olive oil**
- **3/4 cup mushrooms, sliced**
- **2 medium-sized tomatoes, diced**
- **2 cloves garlic, minced**
- **1 small bunch of spinach, chopped**
- **1/2 tsp. soy sauce**
- **1 tsp. lemon juice**

Start off by heating the olive oil, and sautéing the mushrooms, tomatoes and garlic for a few minutes until the garlic starts to get tender. Turn down the heat, and crumble the tofu into the pan along with the remaining ingredients. Keep on cooking for about 5 minutes and you are ready to go.

TIP

Get creative! Look up non-vegan egg recipes and substitute with tofu. **Test out different ingredients and spices to get the flavor right.**

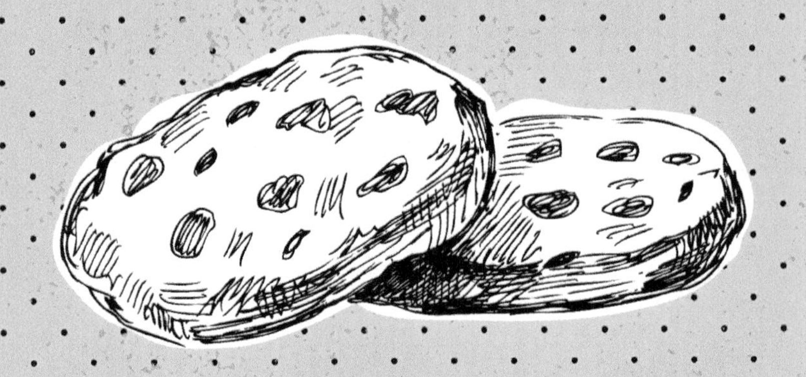

Vegan Chocolate Chip Cookies

TIP

When substituting baking ingredients, make sure to look up the effect it'll have on your final product.

Different sugars will make cookies of different textures.

Vegan Chocolate Chip Cookies

As you probably have figured out, it can be a challenge to bake without milk, eggs or refined sugar. With a few well-chosen substitutes, you can actually still create all the delicious treats you could want. Here is a classic chocolate chip cookie recipe that proves our point.

- **1 cup raw brown sugar**
- **1/2 cup canola oil**
- **6 tbsp. dairy-free milk (soy, rice or almond)**
- **1/4 cup applesauce**
- **2 tsp. vanilla extract**
- **2 1/2 cups all-purpose flour**
- **1 tsp. baking soda**
- **1/2 tsp. salt**
- **1 1/4 cups vegan chocolate chips**

Preheat the oven to 375F and line a couple of baking sheets with parchment.

Beat the sugar, oil, milk, applesauce and vanilla together until everything is smooth and combined. In another bowl, sift flour, baking soda, and salt together. Slowly stir the flour mixture into the first bowl, until it's all mixed. Fold in chocolate chips and pop the entire bowl of dough into the fridge for about an hour.

Once it's chilled, drop spoonfuls of dough onto your baking sheets, to make about 30 to 36 cookies. Back for only about 10 minutes (careful to not over-bake these). Cool on a wire rack and they're ready to eat.

Meatless Tacos

The simplest way to make vegan tacos is to simply use your regular recipe and substitute a vegan meat crumble in place of the ground beef. If you prefer to adjust the recipe to take out the meat component entirely, give these yummy veggie-heavy versions a try.

- **3 cups cabbage, shredded**
- **1 cup onion, diced**
- **1 cup red bell pepper, sliced in strips**
- **2 tbsp. olive oil**
- **1 can black beans, rinsed and drained**
- **1 cup salsa**
- **1 small can of green chilies (chopped)**
- **1 tsp. chili powder**
- **1 tsp. garlic, minced**
- **1/4 tsp. cumin**
- **1/2 cup vegan cheese shreds (optional)**

Heat up oil in a large frying pan, and sauté cabbage, onion and pepper until it all starts to just get tender. Now just stir in everything except for the cheese and let it simmer until the black beans are nicely hot, and the flavors are combined.

Use this filling with either soft tortillas or hard taco shells, topping with a few pinches of vegan cheese if you want.

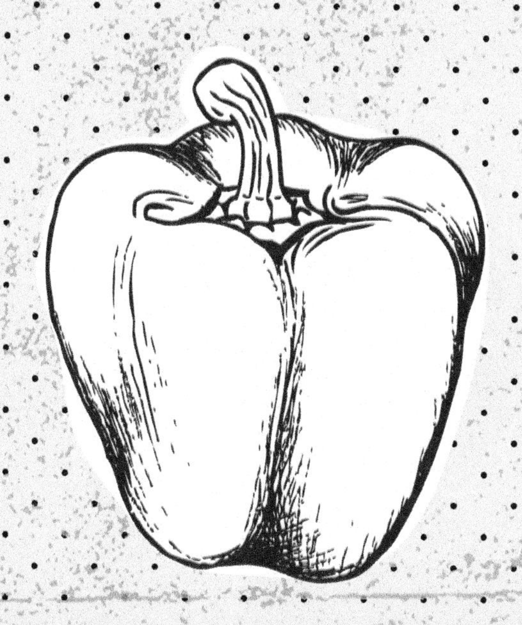

Meatless Tacos

TIP

Add guacamole to your tacos for an extra tasty taco. Most guacamole recipes are vegan, but always double-check.

Apple Spinach Smoothie

TIP

You can make a thicker smoothie
if you freeze the apple and/or avocado before blending.

Apple Spinach Smoothie

Back in the section on eating more greens, we mentioned how you can hide healthy greens in a smoothie. Well, it's only fair that we show you how it's done and this juice-based smoothie is a perfect example. You won't even know you're drinking spinach.

- **1/2 cup water**
- **1/2 cup apple juice**
- **1 tbsp. walnut pieces**
- **1/2 tsp. cinnamon**
- **1 cup cucumber, chopped**
- **2 cups raw spinach, torn up**
- **1 small green apple, chopped**
- **1/4 of an avocado, chopped**
- **5 ice cubes**

You can also adjust the amount of ice to your liking for thickness. Blend everything together until it's smooth and delicious!

Indian Tempeh Curry

Once you're ready to go beyond the basics, here is a very flavorful recipe using some tempeh. You can mix up the vegetable proportions to personalize this Indian dish.

- **2 carrots, sliced or slivered**
- **1 small onion, diced**
- **4 large mushrooms (any variety), sliced**
- **2 potatoes, cubed**
- **1 package of tempeh, sliced**
- **1/2 tbsp. curry powder**
- **1 inch of ginger, grated**
- **2 tbsp. garlic, minced**
- **1 tsp. turmeric**
- **1 can coconut milk**
- **1 1/2 vegan vegetable stock**
- **1 cup fresh cilantro, chopped**
- **1 cup bean sprouts**

First sauté your onions and mushrooms together in oil, then add in the carrots, potatoes, and tempeh until everything is just tender. Set them aside for the moment.

In the bottom of a large stock pot, sauté the garlic, ginger, curry, and turmeric. Once the garlic starts to brown, stir in the coconut milk and vegetable stock. Let it simmer for about 15 to 20 minutes until it begins to thicken up.

Now mix in your batch of vegetables and bring it all back up to a simmer. Keep cooking until the veggies are all cooked and heated through. Serve right away, topped with cilantro and sprouts.

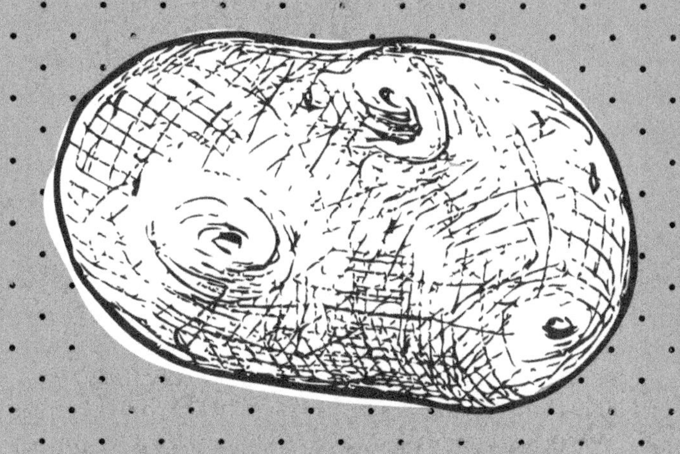

Indian Tempeh Curry

TIP

There are three general **guidelines to making good curry:**

1. Be generous with spices.
2. Either lightly cook or caramelise the onion, ginger, and garlic.
3. Pick an ingredient like tomato, coconut, spinach, or cream to give it body.

Creamy Stuffed Shells

TIP

If you love a cheesy taste in your pasta, remember to add nutritional yeast flakes.

Creamy Stuffed Shells

This is a filling baked pasta dish that uses tofu instead of heavy cream and cheese ingredients. It's a fine main course dish, maybe served with a light green salad.

- **Large pasta shells (around 20 should do)**
- **14 oz. firm tofu**
- **2 cups spinach, raw and chopped**
- **3 cloves garlic, roughly chopped**
- **1/2 tsp. garlic powder**
- **1 jar of your favorite vegan marinara-style sauce**

Start your oven heating to 400F and grease a baking dish large enough to hold your shells when they're laid out.

In boiling water, cook the pasta until they are just barely tender (basically still a little under-cooked). Set that aside and make up the filling by combining the tofu, spinach, and garlic in a food processor and blend until smooth.

Pour some of the marinara in the bottom of your dish, and lay out each cooked shell once you've filled it with the tofu and spinach filling. Pour the rest of your sauce over all the shells, then cover the dish with foil. Add a few small air holes, and bake for about 20 minutes. Spoon out the shells to serve.

Vegan Vanilla Cake

This is a fantastic cake recipe that you can tweak and adapt to suit any occasion. Here's another opportunity to practice your vegan baking skills with a light and tasty cake. You can use this recipe for some basic cupcakes too.

- **1 cup non-dairy milk (unflavored or vanilla)**
- **1 tbsp. vinegar**
- **1 1/2 cups all-purpose flour**
- **1 cup raw vegan sugar**
- **1 tsp. baking soda**
- **1 tsp. baking powder**
- **1/3 cup canola oil**
- **1/4 cup water**
- **1 tbsp. lemon juice**
- **2 tbsp. vanilla extract**

Oven should be preheating to 350F, and you need to grease an 8x8 cake pan.

Combine the vinegar and milk, and set aside. In another mixing bowl, sift together flour, sugar, baking soda, and baking powder. Now add the oil, water, lemon juice and vanilla to the bowl of milk. Give it a good whisking to get the oil nicely combined. Add this liquid mixture to the bowl of flour and stir until you get all the lumps out. Pour into your cake pan.

Bake for about 35 minutes, or until a toothpick comes out cleanly. Frost with any vegan frosting, or even just some fresh fruit.

TIP

For homemade vegan frosting, you only need baked macadamia nuts, shredded unsweetened coconut, vanilla extract, and powdered sugar. Blend the first two separately, then blend them together. Add vanilla and sugar to taste, and blend again.

TIP

For an extra cool treat, toss all the ingredients in a blender instead of a shaker with a good amount of ice and voila! You'll have a frozen margarita.

Vegan Margarita

A lot of pre-made margarita mixes at the store have non-vegan sugar in them. To avoid this problem and cool off with a delicious drink in the summer, here's the vegan version:

- 1 1/2 oz. tequila
- 1 oz. fresh lime juice
- 1/2 oz. orange liqueur (try Cointreau or Triple Sec)
- 1/2 tsp. simple syrup
- Ice
- 1 lime wedge
- Kosher salt

Combine the tequila, lime juice, orange liqueur, simple syrup, and ice in a cocktail shaker and shake well for about 15 seconds.

Salt the margarita glass by rubbing the lime wedge around the rim and then dipping the glass into salt on a plate.

Fill the glass with ice and strain the margarita into the glass. Add a garnish with a lime wedge.

Bonus

7 Vegan Cooking Hacks You Should Know

Vegan Buttermilk:

1. Add **1 table spoon** of **vinegar** to **1 cup** any **non-dairy milk**. Stir with a fork or whisk. **Allow mixture to rest for 5-10 minutes.** Use for making biscuits, pancakes, muffins, and other fluffy baked goods.

Whipped Cream:

2. **Get one can of full-fat vegan coconut milk** (no guar gum in it). **Place the can in the fridge overnight.** Remove can and flip upside down. Open the can and pour the liquid into another bowl or throw it away. **Scoop the remaining coconut cream into a bowl.** Whip the cream until fluffy.

Vegan Donuts:

3. **Get 1 can of biscuit dough** (a few brands are accidentally vegan), 1 medium-sized pot, cooking oil, and frosting or another donut topping. Turn on stove to medium heat. **Pour oil into pot about 1 ½ inches deep.** While the oil is heating, open can of dough and poke a hole into the middle of each biscuit. **When the oil is hot, drop biscuits into oil 2-3 at a time.** Cook until each side is slightly brown. Remove from pot and flavor with powdered sugar or frosting and sprinkles.

Vegan Chocolate Cake:

4. Find a cake mix that's vegan until you add eggs (check the ingredients). **Substitute the eggs with a can of 12 oz. Coca-Cola.** Just mix and bake. Enjoy, you won't taste the soda!

Bacon Flavor:

5. **Some bacon salts are vegan since the flavoring is artificial.** Consume at your own risk, but it might fix your craving for pig fat.

Almond Flour:

6. **If you're making homemade almond milk, don't throw away the pulp!** Strain the mixture with cheese cloth. Then, line a cookie sheet with parchment paper. Place in the oven at the lowest setting for a few hours (about 3). Once it's dry, let it cool. Then toss it into a blender to get a finer texture. **Store in an airtight container until you need to make something delicious, like macaroons!**

Reading Labels:

7. Figuring out what's vegan and what isn't can become a headache. **Apps like Is It Vegan? will make the process easier.** Simply download the app & scan a barcode of store-bought food to check if it's in their database. **The app will let you know which ingredients don't pass the test.**

All Substitutions = 1 Egg

1 tbsp. flaxseed + 3-4 tbsp. water + puree in blender 1-2 minutes

1 tbsp. chia seed + 1/3 cup of water + mix and let sit for 15 minutes

3 tbsp. peanut butter

1 tbsp. soy protein powder + 3 tbsp. water + combine the two

2 tbsp. baking powder + 2 tbsp. water + 1 tbs. oil

3 tbsp. chickpea flour + 3 tbsp. water

½ banana + mash the banana

2 tbsp arrowroot + 3 tbsp water + combine the two

2 tbsp. cornstarch + 3 tbsp. water + combine the two

¼ cup of applesauce

¼ cup silken tofu

Almond Milk: Thick texture, nutritious, and good for cereal and coffee

Hazelnut Milk: Strong flavor and aroma, thin texture, and great in coffee

7-Grain Milk: Thin consistency, nutritious, good for smoothies and cereal

Oat Milk: Thick and grainy, good for hearty cookies

Coconut Milk: Thick, creamy texture, good for moist cakes & with ice cream

Soy Milk: High in protein and fat, creamy texture, versatile uses

Rice Milk: Plain and sweet flavor, watery texture, good for cereal and smoothies

Hemp Milk: High in omega fats and calcium, thick and creamy texture

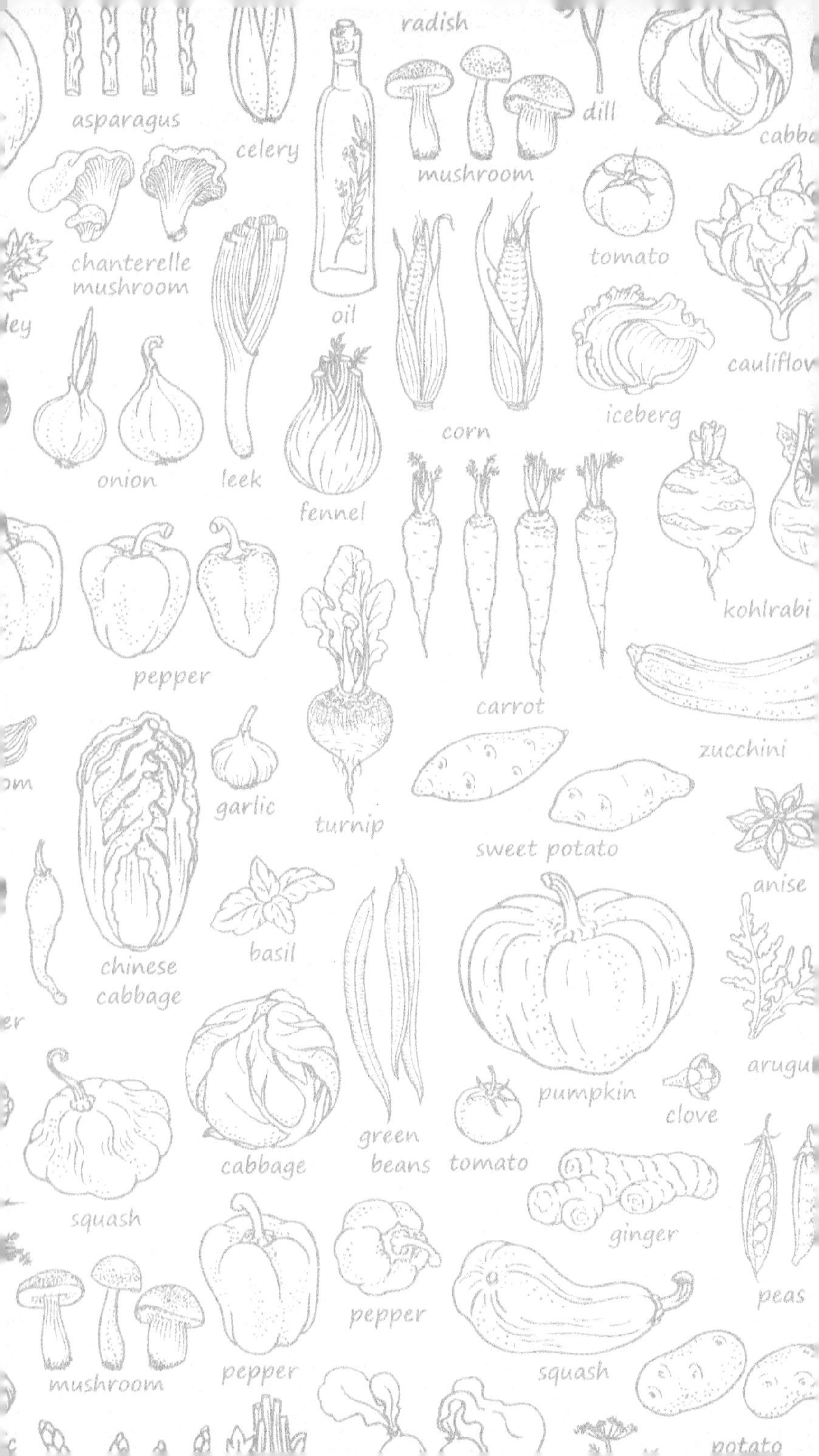

radish
asparagus
celery
mushroom
dill
chanterelle mushroom
tomato
oil
corn
iceberg
onion
leek
fennel
kohlrabi
pepper
carrot
zucchini
garlic
turnip
sweet potato
anise
chinese cabbage
basil
pumpkin
clove
green beans
tomato
cabbage
squash
ginger
peas
pepper
mushroom
pepper
squash
potato

Ideas for Planet-Positive Living

Let's Not Contribute to the "Great Pacific Garbage Patch"

I don't know about you but photos of the big patch of plastic and garbage floating in the ocean scares me more than almost anything else. Nearly 90% of plastic bottles are not recycled, instead taking thousands of years to decompose. If you are used to toting around your green tea, juice or iced coffee in plastic, get a cool-looking thermos instead. This is a great choice for the environment, your wallet, and possibly your health. You can guzzle as much as you want and still be "green."

Save the Rainforest!

Tropical rainforests take in vast quantities of carbon dioxide (a poisonous gas which mammals exhale) and through the process of photosynthesis, converts it into clean, breathable air. In fact, the tropical rainforests are the single greatest terrestrial source of air that we breathe.

What's truly amazing, however, is that while the tropical rainforests cover just 2% of the Earth's land surface, they are home to two-thirds of all the living species on the planet. Additionally, nearly half the medicinal compounds we use every day come from plants endemic to the tropical rainforest. If a cure for cancer or the common cold is to be found, it'll almost certainly come from the tropical rainforests.

Tragically, the tropical rainforests are being destroyed at an alarming rate. According to Rainforest Action Network, more than an acre-and-a-half is lost every second of every day (refer to the entries below to see, quantitatively, what that translates into). That's an area more than twice the size of Florida that goes up in smoke every year!

"If present rates of destruction continue, half our remaining rainforests will be gone by the year 2025, and by 2060 there will be no rainforests remaining." http://www.savetherainforest.org

Cloth Napkins Are Nicer Anyway

On average, an American uses around six napkins each day—2,200 a year! If every American used even one less napkin per day, more than one billion pounds of napkins could be saved from landfills each year.

Reimagine and Reuse Every Chance You Get

In addition to recycling, also reuse. When wrapping presents, use old maps or even newspaper; maybe open up a paper grocery bag, flip it over, and have your kids customize the paper with their artwork. You can also keep and reuse gift bags and tissue paper you were once gifted. This will save you money on buying gift-wrap and help the environment save a few more trees.

Fun in the Sun

Solar ovens are inexpensive to buy, easy to use, and you'll cook for free every time you use one. Solar ovens cook food without using electricity, fossil fuels, or propane. Food cooked in a solar oven retains all vitamins and minerals. Solar ovens also pasteurize water for drinking. A solar oven is perfect for your emergency supply kit. Check out www.solarovens.org to see the great work this nonprofit is doing with solar ovens in Third World countries. Go solar and really worship the sun.

Breathe Easier

This is a great idea for at home and at work. Not only are they lovely to look at, they are improving the air you breathe! These air-purifying plants look great, produce oxygen, and can even absorb contaminants like formaldehyde and benzene (commonly off-gassed from furniture and mattresses). The best part? Nary an electrical cord, nor a battery in sight. Ahhhhh.

- Spider plant
- Peace lily
- Snake plant
- Elephant ear
- Weeping fig
- Rubber plant
- Bamboo palm

Making the Most of a Rainy Day

Get a rain barrel:

- Install the rain barrel at least six feet from your house. Locating it near the area you'll be watering the most makes for convenient use later.

- Ensure that your rain barrel has an overflow at least as large as your inflow—for example, if you have rigged it so that water is collected directly from your eaves' trough downspout, your overflow valve should be as large as your downspout as well. This will allow your barrel to get rid of excess water as fast as it collects it, which might be necessary if you live in a city with crazy, unpredictable weather like my brother does.

- If you are using the rain barrel to water your garden, consider using a soaker hose. You can attach the hose to the rain barrel and then run it through your garden so that it covers the area you'd like. Now, every time you see a rain cloud, you'll get really excited!

More Ways to Avoid Plastic

- Jars. So many jars. For salads, soups, sauces, juice, opened packages of things. Just jars. Lots of jars.

- Aluminum foil is a great stand-in for plastic wrap and can be easily recycled or saved for reuse.

- Wax paper is great for wrapping sandwiches. So are reusable sandwich pouches you can make or purchase; many close with Velcro and are easy to launder or wipe clean.

- Reuse empty yogurt, sour cream, or cream cheese containers. You've already got them and you have to wash them before tossing them into the recycling bin anyway, so you may as well use them again.

- Invest in reusable lunch containers like bento boxes or tiffins to avoid waste when sending off your little one (or yourself) with a bagged lunch.

How to be "Water Wise"

- One cup of water costs five times as much in a Nairobi slum as in an American city.

- Three gallons of water provide the daily drinking, washing, and cooking water of one person in the developing world...yet in the U.S. it flushes one toilet.

- Three gallons of water weigh 25 pounds. Women in Africa and Asia carry, on average, twice this amount of water over 4 miles...each and every day.

- 470 million people live in regions of severe water shortage. It's estimated, if nothing is done, that by the year 2025 this will increase six fold.

- Roughly 1/6 of our World's population does not have access to safe water.

- 2 ½ billion people (roughly 1/6 of the world's population) do not have adequate sanitation and according to the United Nations, 6,000 children each day die from unsafe water and sanitation: that is the equivalent of 20 jumbo jets filled with children crashing every day.

Further Readings

http://www.npr.org/blogsmoney/2012/06/08/154568945/what-america-spends-on-groceries

http://www.vegansociety.com/your-business/vegan-trademark-standards

http://www.vrg.org/blog/2011/12/05/how-many-adults-are-vegan-in-the-u-s/

http://darwin.bio.uci.edu/sustain/global/sensem/MeatIndustry.html

http://www.peta.org/issues/animals-used-for-food/meat-wastes-natural-resources/

http://www.nhlbi.nih.gov/health/health-topics/topics/hbc

http://www.ucsusa.org/food_and_agriculture/our-failing-food-system/industrial-agriculture/prescription-for-trouble.html

http://www.dairydoingmore.org/economicimpact/statistics/dairystatistics

http://www.npr.org/blogs/thesalt/2011/12/31/144478009/the-average-american-ate-literally-a-ton-this-year

http://www.nationalchickencouncil.org/about-the-industry/statistics/per-capita-consumption-of-poultry-and-livestock-1965-to-estimated-2012-in-pounds/

http://www.aeb.org/farmers-and-marketers/industry-overview

http://www.honey.com/newsroom/press-kits/honey-industry-facts

http://www.iwto.org/about-iwto/vision-mission/

http://ods.od.nih.gov/factsheets/Calcium-HealthProfessional/

http://greatist.com/health/18-surprising-dairy-free-sources-calcium
http://www.vrg.org/nutrition/iron.php

http://eating-mindfully.com/black-bean-burgers

http://www.thekitchn.com/vegetarian-lunch-chickpea-of-t-114022

http://vegetarian.about.com/od/breakfastrecipe1/r/spinachscramble.htm

http://www.tasteofhome.com/recipes/vegan-chocolate-chip-cookies

http://www.tasteofhome.com/recipes/veggie-tacos

http://www.healthfulpursuit.com/recipe/apple-pie-green-smoothie/

http://foyupdate.blogspot.ca/2010/02/vegetarian-tempeh-curry-indian-recipe.html

http://www.popsugar.com/fitness/Vegan-Stuffed-Shells-36258405/

http://allrecipes.com/recipe/vegan-basic-vanilla-cake/

https://www.vegansociety.com/sites/default/files/uploads/trademark-logo.png

https://veganfoodaddict.wordpress.com/2011/11/06/vegan-buttermilk-did-you-know/

http://allsortsofpretty.com/how-to-make-almond-milk-and-almond-flour-a-surprisingly-easy-diy/

http://www.buzzfeed.com/leonoraepstein/hacks-every-vegan-should-know?sub=3191089_2839111#.yc8YMqY3d0

radish
asparagus
celery
dill
mushroom
chanterelle mushroom
tomato
oil
corn
iceberg
onion
leek
fennel
kohlrabi
pepper
carrot
zucchini
garlic
turnip
sweet potato
anise
chinese cabbage
basil
pumpkin
clove
cabbage
green beans
tomato
squash
ginger
peas
pepper
mushroom
pepper
squash

GOOD FOR YOU; GOOD FOR YOUR WALLET

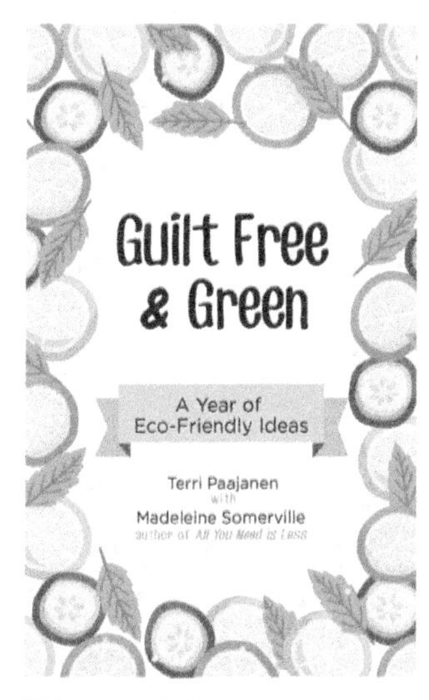

ISBN 978-1-63353-137-6
PRICE $16.95
TRIM 5x8

ISBN 978-1-63353-654-8
PRICE $14.95
TRIM 5x8

ISBN 978-1-63353-657-9
PRICE $16.95
TRIM 5x8

ISBN 978-1-63353-660-9
PRICE $14.95
TRIM 5x8

Thank you for reading.

In writing *Vegans Save the World: Plant-based Recipes and Inspired Ideas for Every Week of the Year*, Alicia Alvrez did her very best to produce the most accurate, well-written and mistake-free book. Yet, as with all things human (and certainly with books), mistakes are inevitable. Despite the publisher's best efforts at proofreading and editing, some number of errors will emerge as the book is read by more and more people.

We ask for your help in producing a more perfect book by sending us any errors you discover at errata@mango.bz. We will strive to correct these errors in future editions of this book. Thank you in advance for your help.

CPSIA information can be obtained
at www.ICGtesting.com
Printed in the USA
BVOW10s0039181017
497950BV00003B/4/P